Handbook of Liquefied
Natural Gas

Handbook of Liquefied Natural Gas

Editor

Deependra Singh

Handbook of Liquefied Natural Gas

Edited by **Deependra Singh**

Printed in 2017

ISBN: 978-1-68117-393-1

Library of Congress Control Number: 2015941582

© 2016 by

SCITUS Academics LLC,
616, Corporate Way, Suite 2, 4766,
Valley Cottage, NY 10989

www.scitusacademics.com

Contents

Preface

Liquefied natural gas (LNG) is a commercially attractive phase of the commodity that facilitates the efficient handling and transportation of natural gas around the world. The LNG industry, using technologies proven over decades of development, continues to expand its markets, diversify its supply chains and increase its share of the globa natural gas trade. The Handbook of Liquefied Natural Gas is a timely book as the industry is currently developing new large sources of supply and the technologies have evolved in recent years to enable offshore infrastructure to develop and handle resources in more remote and harsher environments. This book provides an ideal platform for scientists, engineers, and other professionals involved in the LNG industry to gain a better understanding of the key basic and advanced topics relevant to LNG projects in operation and/or in planning and development. The LNG supply chain extends from upstream production, LNG production plant, shipping, storage, and regasification to supply to sales gas pipe ines and power plant users. LNG production is capital intensive and the recent costs have deterred the commitment of most investors, and any future LNG production plant owners must reevaluate the current technologies for a -fit-for-purpose" design to reduce the life cycle costs.

Editor

Hydrogen Production through the Fuel Processing of Liquefied Natural Gas with Silicon-based Micro-reactors

Sang Hyun Ahn[a], Insoo Choi[a], Oh Joong Kwon[b], and Jae Jeong Kim[c]

[a]Fuel Cell Research Center, Korea Institute of Science and Technology, Hwarangno 14-gil 5, Seongbuk-gu, Seoul 136-791, Republic of Korea

[b]Department of Energy and Chemical Engineering, Incheon National University, 12-1, Songdo-dong, Yeonsu-gu, Incheon 406-840, Republic of Korea

cSchool of Chemical and Biological Engineering, Institute of Chemical Process, Seoul National University, Gwanak-ro 1, Gwanak-gu, Seoul 151-742, Republic of Korea

ABSTRACT

Silicon-based micro-reactors for the fuel processing of liquefied natural gas (LNG) were fabricated using silicon technologies. The micro-LNG steam reformer achieves a LNG conversion of 77.4%, and the hydrogen composition of the product was 73.3% at 600°C. The product gas was supplied to consecutive micro-reactors to carry out carbon monoxide removal through a high-temperature water–gas shift (HTS) reaction and a low-temperature water–gas shift (LTS) reaction. Under operating conditions we investigated, the micro-HTS and LTS reactors demonstrated the highest carbon monoxide conversion of 61.5% at 450°C and 77.5% at 300°C, respectively. The final product gas of the micro-fuel processor was composed of 75.7% hydrogen and 0.7% carbon monoxide.

INTRODUCTION

Hydrogen has been considered as an attractive energy source owing to its high energy efficiency and environmentally friendly characteristics [1] and [2]. Therefore, several hydrogen production technologies, such as hydrocarbon steam reforming, coal gasification, enzymatic hydrogen generation and electrolysis (e.g., photocatalytic, photobiological and biocatalysed), have been studied to prepare for the high demand for hydrogen in the near future [3] and [4]. Among the various hydrogen production methods which have been developed, hydrocarbon steam reforming is considered as a highly feasible method because it shows the highest energy efficiency in generating hydrogen, despite the fact that it produces greenhouse gases during the reforming process [5] and [6]. In particular, LNG consisting of high methane (~92%) has been recognized as a suitable hydrocarbon to produce hydrogen with a low level of energy consumption because methane has a high hydrogen-to-carbon ratio (CH_4) compared to other hydrocarbons [7], [8] and [9]. The LNG infrastructure is also well established, thus facilitating the production of hydrogen from LNG with a residential reformer [10]. In addition, the LNG is a clean energy source because it does not contain dust, sulfur or nitrogen, which are main causes of environmental pollution.

In order to achieve reasonable conversion via the steam reforming of methane (SRM), it was originally considered that SRM requires a high operating temperature of over 700 °C, a high steam-to-carbon (S/C) ratio close to 3.0, and a highly pressurized reactant at over 20 bar [11], [12] and [13]. These severe conditions limited the SRM process as a hydrogen production method for several decades. However, with the advances in catalyst technology and the optimization of operating conditions, recent studies have reported that SRM can be carried out at a relatively low operating temperature (~600 °C), at a low S/C ratio around 2.0, and under atmospheric pressure [14], [15] and [16].

To obtain high-quality hydrogen through LNG steam reforming, the content of carbon monoxide in the product gas should be reduced or eliminated because low-quality hydrogen limits its application. For example, the performance of especially fuel cells drops when used with hydrogen containing carbon monoxide. The platinum catalyst in the fuel cell is poisoned by the strong adsorption of carbon monoxide[17] and [18]. To lower the carbon monoxide concentration, the water–gas shift (WGS) reaction ($CO + H_2O \leftrightarrow H_2 + CO_2$), which not only removes carbon monoxide in the product gas from LNG steam reformers but also produces additional hydrogen, has been investigated by many researchers [19] and [20], though it has reached a technical plateau.

Many studies have reported fuel processors with micro-reactor operation due to their high efficiency, fast heat transfer rates, high surface-to-volume ratios, and simple scale up processes [21], [22], [23], [24], [25], [26],[27], [28], [29], [30], [31], [32], [33] and [34]. We also reported a methanol fuel processor based on a silicon-based micro-reactor composed of a methanol steam reformer and a preferential oxidation reactor (PrOx)[35], [36], [37], [38] and [39]. In that study, the methanol steam reformer achieved 95% methanol conversion with a carbon monoxide concentration of 2100 ppm at 320 °C. The PrOx reactor decreased the carbon monoxide concentration in the product gas such that it was lower than the detection limit of our gas chromatograph (GC) at 220 °C. However, research on LNG steam reforming during the operation of a micro-reactor is relatively rare because the operating temperatures required during the LNG steam reforming process are much higher than those needed during methanol and ethanol steam reforming, although the commercialization of the process has the advantages due to the well-established infrastructure, as mentioned above [30], [31] and [32].

In this study, the application of a silicon-based micro-reactor is broadened to the operation of a LNG fuel processor consisting of a LNG steam reformer for hydrogen production and HTS/LTS reactors for carbon monoxide removal. After the fabrication processes, the performance of the micro-LNG steam reformer was tested under various operating conditions and then compared with the result of a conventional continuous flow fixed-bed reactor. The product gas from the micro-LNG steam reformer was passed through the micro-HTS and LTS reactors in sequence, and the composition of the exhaust gas was then compared with that exhausted from conventional HTS and LTS reactors.

EXPERIMENTAL

The detailed fabrication process of micro-reactor is described in our previous work [35]. Thus, the fabrication process is briefly introduced in this study. It is also summarized in Fig. 1. The steps were the preparation of a Si (1 1 0) wafer (Fig. 1a), the formation of a micro-channel with photolithography and chemical wet etching (Fig. 1b), the isolation of the micro-channel using anodic bonding followed by a hole fabrication step (Fig. 1c), the deposition of a Ta/TaNx thin film heater. In this experiment, an improvement on Ta/

TaNx thin film heater was made, that is, an SiNx oxygen barrier was additionally deposited via physical vapor deposition (PVD) for preventing the degradation of thin film heater above 400 °C (Fig. 1d). Following processes were the coating of the catalyst layer on the inner wall of the micro-channel (Fig. 1e), and finally packaging with stainless steel materials (Fig. 1f).

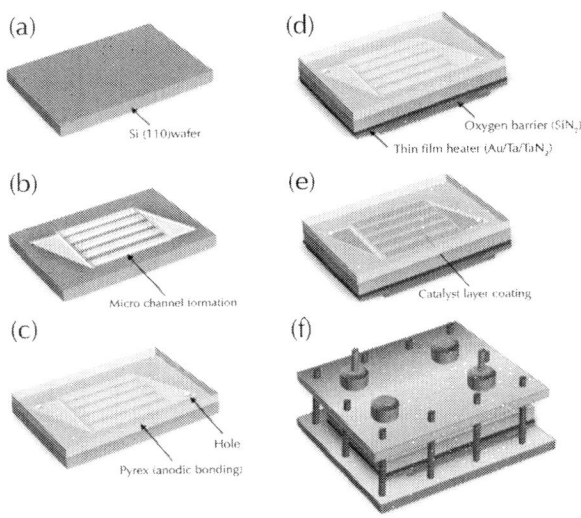

Figure 1: Schematic diagram of the micro-reactor fabrication process: (a) preparation of the Si (1 1 0) wafer, (b) formation of the micro-channel, (c) isolation of the micro-channel, (d) deposition of the thin film heater including the oxygen barrier, and (e) coating of the catalyst layer and (f) packaging of the micro-reactor.

For the catalyst coating step (Fig. 1e), a lab-made mesoporous Ni–Al$_2$O$_3$ composite catalyst was used for LNG steam reforming, and the catalyst preparation steps were described in the reported literature [40]. For both of the WGS reactions, a commercial Pt/Fe$_2$O$_3$/Cr$_2$O$_3$/CeO$_2$/CuO (HiFuel™ W210) catalyst for the HTS reaction and CuO/ZnO/Al$_2$O$_3$ (HiFuel™ W230) as a catalyst for the LTS reaction were purchased from Alfa Aesar. Catalyst powder was mixed with deionized water to prepare a catalyst slurry, which was then coated onto the micro-channel using a fill-and-drying method

[35]. Prior to the catalyst coating process, all of the catalysts were activated by reduction for 3 h in a passing gas mixture of 4 vol% hydrogen in balanced nitrogen at 700 °C for LNG steam reforming, at 600 °C for HTS, and 350 °C for LTS. The catalyst coating profile was observed by field-emission scanning electron microscopy (FESEM, S-4800; Hitachi). The fabricated micro-reactor was packaged with stainless steel and a metal O-ring to test its performance for LNG steam reforming and WGS reactions (Fig. 1f).

Fig. 2 illustrates a schematic diagram of the experimental setup to test the performance of the micro-reactors. The temperature of the micro-reactor was controlled by a power supply, a relay (RY5 W-K; LG) and a proportional-integrate-derivative (PID) algorithm programmed by LabVIEW™. Deionized water was vaporized by heating a coil (120 °C). The vapor was then mixed with LNG by feeding it into the micro-LNG steam reformer. The feed rates of the deionized water and gases were controlled by a syringe pump (780200; KD Scientific) and a mass flow controller (MFC, F201CV; Bronkhorst High-Tech), respectively. The feed rate was precisely controlled to fix the GHSV of 27,000 mL/h g_{cat}. After the reactions in the micro-reactors, the steam in the product gas was removed by a cold trap. The product gas was analyzed using an on-line GC (ACME 6000, Younglin).

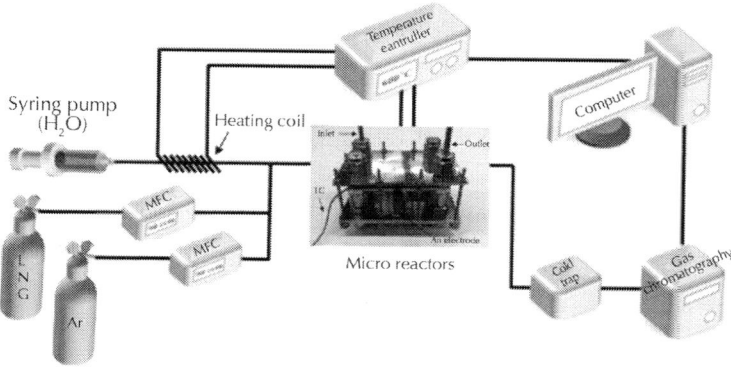

Figure 2: Experimental setup for the performance test of the micro-LNG fuel processor.

For the comparison of the performance of micro-reactor, the conventional continuous flow fixed-bed reactor was adopted. The vertical quartz with a volume of 2.8 mL was used for conventional continuous flow fixed-bed reactor system. The loading mass of each catalyst for LNG steam reforming, HTS, and LTS was 100 mg. The temperature of reactor was controlled by external furnace under atmospheric pressure, and total feed rate was fixed at gas hourly space velocity (GHSV) of 27,000 mL/h g_{cat}. The product gas was periodically sampled and analyzed using an on-line GC (ACME 6000, Younglin).

RESULTS AND DISCUSSION

Fabrication of Micro-reactors for LNG Fuel Processing

Fig. 3a shows the front side of the fabricated micro-reactor. The chemically wet-etched micro-channels are designed as six serpentine channels in parallel with each other. The total length of the micro-channel is approximately 178 cm. The holes are positioned at the top left and the bottom right of the inlet and outlet, respectively. Fig. 3b displays a cross-sectional FESEM image of a rectangular micro-channel with a width of 600 μm and a depth of 240 μm. The volume of the micro-channel in the micro-reactor is 0.26 mL. Fig. 3c illustrates the back side of the micro-reactor after the formation of the thin film heater. The top and bottom Au pad lines were deposited on the Ta/TaNx film for a connection with an external electrical power source. At a high temperature over 400 °C, the severe oxidation of thin film heater was observed and it hindered the precise temperature control of micro-reactor. Thus, for steam reforming of LNG, the other parts of the thin film heater excepting Au pad were covered by the deposition of SiNx (1.7 μm) oxygen barrier to prevent the degradation of the thin film heater. Its cross-section was observed by FESEM to confirm the deposition of the thin film heater, as shown in Fig. 3d.

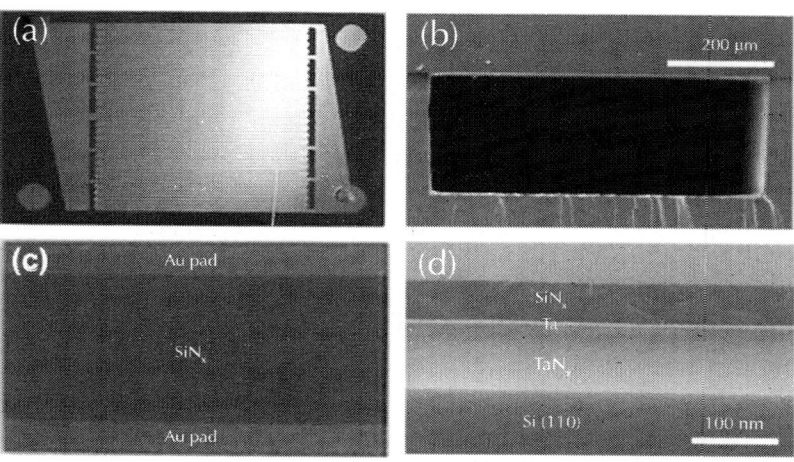

Figure 3: (a) Image of the micro-reactor (front side view), (b) FESEM image of the cross-section of the micro-channel, (c) image of the micro-reactor (back side view), and (d) cross-sectional FESEM image of the thin film heater.

Fig. 4 shows images of the micro-reactor with a cross-sectional image of FESEM (insets) after the catalyst coating step. For catalyst coating in the LNG steam reformer, a lab-made mesoporous Ni–Al_2O_3 composite catalyst was mixed with deionized water to prepare a catalyst slurry having a weight ratio of 20 wt.%. In order to increase the adhesion between the catalyst layer and the inner wall of the micro-channel, acrylic acid (2 wt.%), which controls the viscosity of the catalyst slurry, was added as a binder [41]. The thickness of the coated Ni–Al_2O_3 catalyst layer was 11.2 µm at the top of the channel and 28.6 µm at the side wall of the channel, as illustrated in the inset of Fig. 4a. For a WGS reactor, Pt- and CuO-based commercial catalyst slurries (20 wt.%) were used. In these cases, the catalyst layers were well coated onto the inner walls of the micro-channels without a binder, as shown in inset of Fig. 4b and c. The layer thicknesses of the HTS and LTS reactors were 14.4/37.3 µm (top/side) and 10.6/22.1 µm (top/side), respectively.

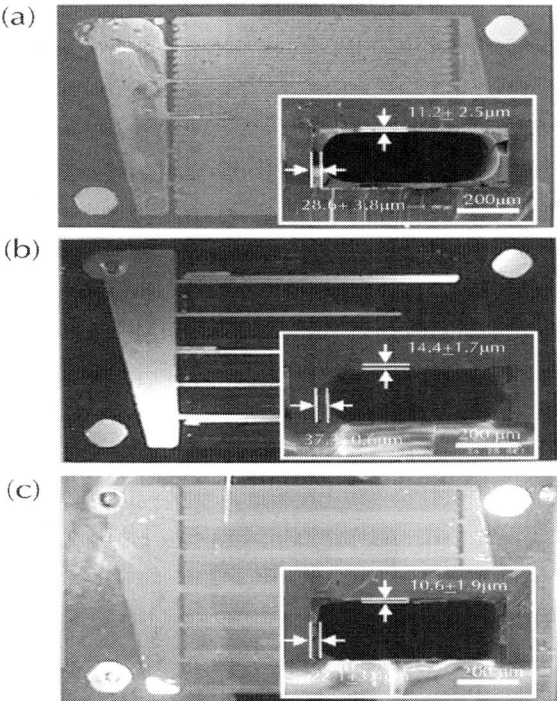

Figure 4: Images of the micro-reactor after the coating of the catalyst layer (front side view) for (a) the LNG steam reformer, (b) the HTS reactor and (c) the LTS reactor. Insets: FESEM images of the cross-section of the micro-channel after the coating of the catalyst layer.

After the deposition of the thin film heater and the coating of the catalyst layer, the micro-reactor was packaged with stainless steel materials and a metal O-ring to connect it to the external instruments (Fig. 2). Before the testing of the performance of the micro-reactor for various reactions, the temperature variation of the micro-reactor was monitored by adjusting the electrical power to confirm the accuracy of the thin film heater, as shown in Fig. 5. Without SiNx oxygen barrier, precise temperature control of micro-reactor was not possible over 400 °C. On the other hand, with the oxygen barrier, the reactor temperature increased linearly by increasing the set temperature to nearly 650 °C. This indicated that the oxygen barrier prevented the degradation of thin film heater at high temperature which required for LNG steam reforming.

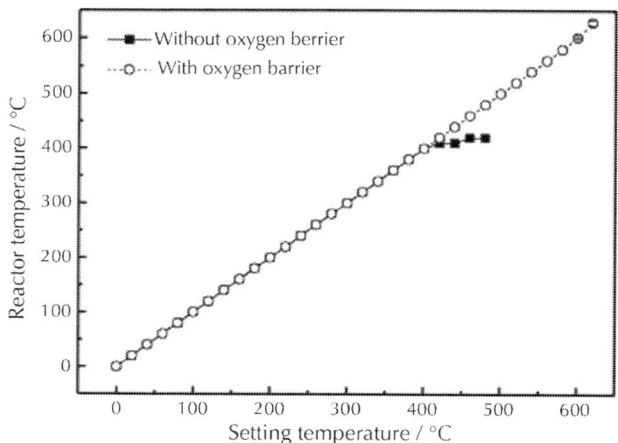

Figure 5: A graph showing the performance of the thin film heater without and with SiNx oxygen barrier.

Performance of Micro-reactors for LNG Steam Reforming

To investigate the effectiveness of the micro-reactors for LNG fuel processing, the reaction was conducted in the micro-reactors as well as in a conventional fixed-bed reactor. For the steam reforming of LNG with the conventional reactor, the total feed rate with respect to the catalyst loading mass was fixed at a GHSV of 27,000 mL/h g_{cat}. The composition of the feed was CH_4:C_2H_6:Ar = 9.2:0.8:90.0 with an S/C ratio of 2.0 [40]. The LNG conversion and hydrogen composition were calculated by the following equations on the basis of the carbon balance [40].

$$\text{LNG conversion}/\% = \left(1 - \frac{F_{CH_4,out} + F_{C_2H_6,out}}{F_{CH_4,in} + F_{C_2H_6,in}}\right) \times 100$$

(1)

Composition of [A] in dry gas/%

$$= \left(\frac{F_{[A],out}}{F_{H_2,out} + F_{CH_4,out} + F_{C_2H_6,out} + F_{CO,out} + F_{CO_2,out}}\right) \times 100$$

(2)

The LNG conversion and hydrogen compositions with a conventional continuous flow reactor at 600 °C were 69.8% and 67.0%, respectively [40].

With the same GHSV with a conventional reactor, the steam reforming of LNG was performed in the micro-reactor at temperatures between 450 and 600 °C. As shown in Fig. 6a, the LNG conversion with the micro-reactor was increased with an increase in the operating temperature, reaching 77.4% at 600 °C. In the operating temperature range, the micro-LNG steam reformer always showed a higher LNG conversion compared to the result of the conventional reactor. Fig. 6b describes the composition of the product gases from the micro-LNG steam reformer. The hydrogen composition of the product gases varied from 49.5% at 450 °C to 73.3% at 600 °C. The carbon monoxide content, which is important to obtain high-quality hydrogen, was enriched from 4.8% to 8.8% as the operating temperature increased.

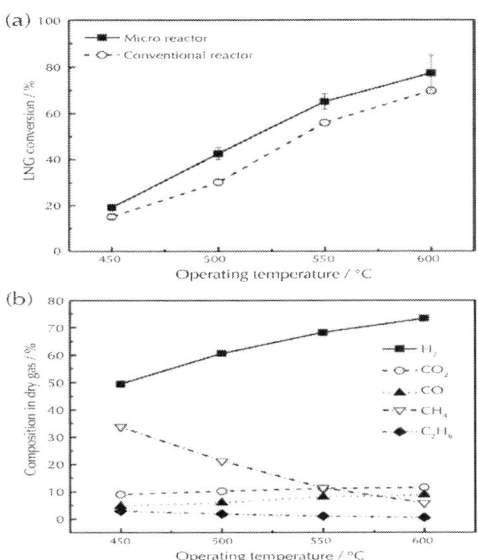

Figure 6: (a) LNG conversion and (b) composition of dry product gas from the LNG steam reformer depending on the operating temperature. Feed composition: CH_4 (9.2%), C_2H_6 (0.8%) and Ar (90.0%) with an S/C ratio of 2.0.

Given that methane steam reforming over Ni catalyst (Eq. (3)) is generally carried out accompanying with secondary reactions; WGS (Eq. (4)), carbon deposition on the Ni catalyst surface (Eq. (5)), carbon gasification (Eq. (6)), and syngas production (Eq. (7)) as described in below [42], [43], [44], [45] and [46]:

$$CH_4 + H_2O \leftrightarrow CO + 3H_2, \quad H_2/CO = 3 \tag{3}$$

$$CO + H_2O \leftrightarrow CO_2 + H_2 \tag{4}$$

$$CH_4 \leftrightarrow C + 2H_2 \tag{5}$$

$$C + H_2O \leftrightarrow CO + H_2 \tag{6}$$

$$CH_4 + CO_2 \leftrightarrow 2CO + 2H_2 \tag{7}$$

Since the LNG feed contained the ethane (~8%), it should be noted that the steam reforming of ethane could be carried out by following equations [47]:

$$C_2H_6 + H_2O \leftrightarrow 2CO + 5H_2, \quad H_2/CO = 2.5 \tag{8}$$

$$C_2H_6 + H_2 \leftrightarrow 2CH_4 \tag{9}$$

Eq. (8) demonstrates the ethane steam reforming and following WGS (Eq. (4)) allows the generation of additional hydrogen, while the methane can be produced by hydrogenolysis (Eq. (9)). Focusing on the main reactant (CH_4: ~92%) in considering carbon monoxide content, the methane steam reforming (Eq. (3)) and WGS reaction (Eq. (4)) are both reversible reactions which can be expressed with a forward and backward reaction rate as a strong function of temperature [48]. With competition between two reactions, it is expected that the exothermic WGS reaction approached the thermodynamics limitation with the increase in operating temperature. The gradual increase in the carbon monoxide content was observed, evidently indicating that the reverse WGS (RWGS) could be promoted at high temperature. In addition, the production of syngas (Eq.(7)) should be considered as one of the possible reasons for the increase in carbon monoxide content [46]. In the previous study [40], it has been confirmed that the carbon deposition on the catalyst (Eq. (5)) and following carbon gasification (Eq. (6)) could be negligible (C content: 0.3 wt.%).

Performance of Micro-reactors for WGS Reactions

The product gas obtained with the micro-LNG steam reformer at 600 °C was supplied to both the micro- and the conventional HTS reactor operated in a temperature range between 300 °C and 500 °C. The recorded steam-to-carbon-monoxide ratio in the product gas of the reformer was 3.7. As shown in Fig. 7a, in the operating temperature range of 300–450 °C, the carbon monoxide conversion in the micro-HTS reactor increased from 21.7% to 61.5%, after which the conversion decreased to 57.8% at 500 °C. The decrease in the conversion at temperatures exceeding 450 °C can be explained based on the exothermic characteristic of the WGS reaction. Fig. 7b demonstrates the compositional variation of the product gas depending on the operating temperature. At 450 °C, the carbon monoxide composition was 3.2% with the highest carbon monoxide conversion in the micro-HTS reactor. In the case of methane, the concentration decreased slightly (5.8% → 5.1%) because the reverse SRM ($CO + 3H_2 \leftrightarrow CH_4 + H_2O$) known as one of the side reactions under WGS reaction could be carried out [49] and [50].

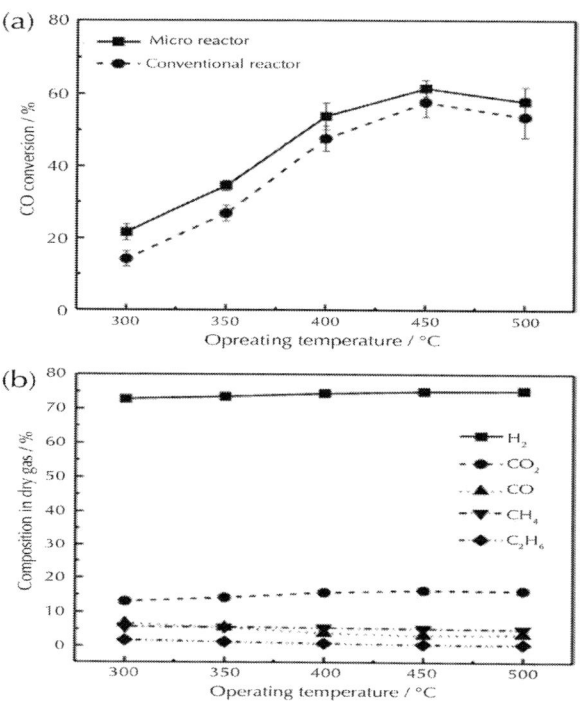

Figure 7: (a) Carbon monoxide conversion and (b) composition of dry product gas from the HTS reactor depending on the operating temperature. Feed composition: H_2 (73.3%), CO_2 (11.6%), CO (8.8%), CH_4 (5.8%) and C_2H_6 (0.5%).

The micro- or conventional LTS reactor was serially connected to the micro-HTS reactor operated at 450 °C. The feed composition injected into the LTS reactors was estimated to be H_2 (75.0%), CO_2 (16.2%), CO (3.2%), CH_4 (5.1%) and C_2H_6 (0.5%) based on the composition of the product gas from the micro-HTS reactor. A copper metal surface is commonly recognized as the primary reaction center in the catalysis of WGS and RWGS reactions at relatively low temperature [51]. Fig. 8a shows the carbon monoxide conversion for both the micro- and conventional LTS reactors as a function of the operating temperature. The conversion of the micro-LTS reactor increased from 41.7% to 77.5% and then became mostly saturated under the investigated temperature range.

In all of temperature ranges, the micro-LTS reactor demonstrated higher carbon monoxide conversion compared to those of the conventional LTS reactor. For example, the carbon monoxide conversion was 5.9% higher in the micro-LTS reactor than in the conventional LTS reactor at 300 °C, at which the micro-LTS reactor showed the highest conversion. As demonstrated in Fig. 8b, the lowest carbon monoxide composition of 0.7% could be obtained at 300 °C, at which the highest conversion was demonstrated. 87.5% of the carbon monoxide was eliminated in the micro-LTS reactor, indicating that the product gas can be directly applied to a high-temperature polymer electrolyte membrane fuel cell (PEMFC) made with a polybenzimidazole (PBI) membrane [52]. This demonstrates that micro-reactors with an optimized catalyst can generate high-quality hydrogen through LNG reforming.

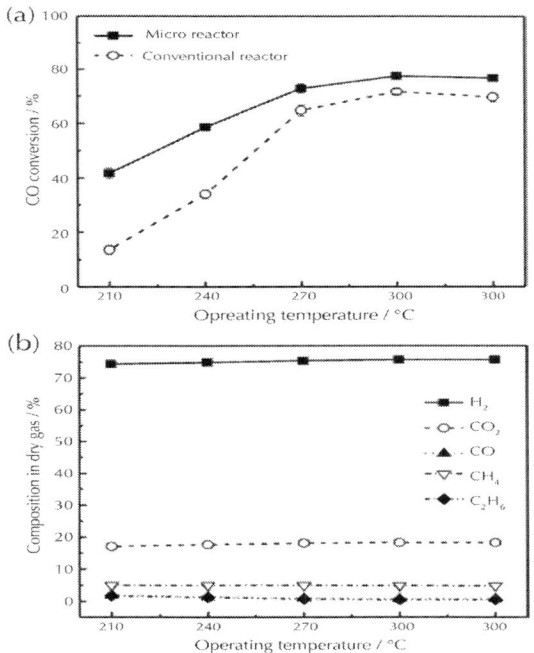

Figure 8: (a) Carbon monoxide conversion and (b) composition of dry product gas from the LTS reactor depending on the operating temperature. H_2 (75.0%), CO_2 (16.2%), CO (3.2%), CH_4 (5.1%) and C_2H_6 (0.5%).

CONCLUSIONS

A micro-LNG fuel processor including a LNG steam reformer and carbon monoxide removal units was successfully fabricated by silicon technology. With a layer-type catalyst and a thin film heater, the micro-LNG steam reformer achieved a LNG conversion of 77.4% at 600 °C with a hydrogen composition of 73.3% and a carbon monoxide of 8.8% in the product gas. This high carbon monoxide composition was decreased to 3.2% with a micro-HTS reactor and was reduced further to 0.7% with a micro-LTS reactor. The hydrogen composition of the final product gas was 75.7%. It is expected that the final product gas can be directly applied to an energy system which requires high-quality hydrogen.

ACKNOWLEDGMENTS

This work was supported by the Basic Science Research Program through the National Research Foundation of Korea (NRF), and funded by the Ministry of Education, Science and Technology (NRF-2010-0029071). This work (Grants No. C0143059) was supported by Business for Cooperative R&D between Industry, Academy, and Research Institute funded Korea Small and Medium Business Administration in 2013.

REFERENCES

1.   P. Barbaro, C. Bianchini, Catalysis for Sustainable Energy Production, WileyVCH, Weinheim, 2009.

2.   W. Vielstich, A. Lamm, H. Gasteiger, Handbook of Fuel Cells: Fundamentals, Technology and Applications, Wiley, Chichester, 2003.

3.   R.M. Navarro, M.A. Peña, J.L.G. Fierro, Hydrogen production reactions from carbon feedstocks: fossil fuels and biomass, Chem. Rev. 107 (2007) 3952–3991.

4.   K. Christopher, R. Dimitrios, A review on exergy comparison of hydrogen production methods from renewable energy sources, Energy Environ. Sci. 5 (2012) 6640–6651.

5.   J.D. Holladay, J. Hu, D.L. King, Y. Wang, An overview of hydrogen production technologies, Catal. Today 139 (2009) 244–260.

6.   S. Ahmed, M. Krumpelt, Hydrogen from hydrocarbon fuels for fuel cells, Int. J. Hydrogen Energy 26 (2001) 291–301.

7.   S.S. Maluf, E.M. Assaf, Ni catalysts with Mo promoter for methane steam reforming, Fuel 88 (2009) 1547–1553.

8.   A.P. Simpson, A.E. Lutz, Exergy analysis of hydrogen production via steam methane reforming, Int. J. Hydrogen Energy 32 (2009) 4811–4820.

9.   S.H. Park, B.H. Chun, S.H. Kim, Effects of La2O3 on ZrO2 supported Ni catalysts for autothermal reforming of CH4, Korean J. Chem. Eng. 28 (2011) 402–408.

10.   Y.T. Seo, D.J. Seo, J.H. Jeong, W.L. Yoon, Development of compact fuel processor for 2 kW class residential PEMFCs, J. Power Sources 163 (2006) 119–124.

11.   J.N. Armor, The multiple roles for catalysis in the production of H2, Appl. Catal. A: Gen. 176 (1999) 159–176.

12.   J.R. Rostrup-Nielsen, Sulfur-passivated nickel catalysts for carbon-free steam reforming of methane, J. Catal. 85 (1984) 31–43.

13.   K. Koo, J. Yoon, C. Lee, H. Joo, Autothermal reforming of methane to syngas with palladium catalysts and an electric metal monolith heater, Korean J. Chem. Eng. 25 (2008) 1054–1059.

14.   Y. Bang, S.J. Han, J. Yoo, J.H. Choi, K.H. Kang, J.H. Song, J.G. Seo, J.C. Jung, I.K. Song, Hydrogen production by steam reforming of liquefied natural gas (LNG) over trimethylbenzeneassisted ordered mesoporous nickelealumina catalyst, Int. J. Hydrogen Energy 38 (2013) 8751–8758.

15. Y. Bang, J.G. Seo, M.H. Youn, I.K. Song, Hydrogen production by steam reforming of liquefied natural gas (LNG) over mesoporous Ni–Al2O3 aerogel catalyst prepared by a single-step epoxide-driven sol–gel method, Int. J. Hydrogen Energy 37 (2012) 1436–1443.

16. Y. Bang, J.G. Seo, I.K. Song, Hydrogen production by steam reforming of liquefied natural gas (LNG) over mesoporous Ni–La–Al2O3 aerogel catalysts: effect of La content, Int. J. Hydrogen Energy 36 (2011) 8307–8315.

17. J.C. Amphlett, R.F. Mann, B.A. Peppley, On board hydrogen purification for steam reformation/PEM fuel cell vehicle power plants, Int. J. Hydrogen Energy 21 (1996) 673–678.

18. M. Kim, B. Fang, N.K. Chaudhari, M. Song, T. Bae, J. Yu, A highly efficient synthesis approach of supported Pt–Ru catalyst for direct methanol fuel cell, Electrochim. Acta 55 (2010) 4543–4550.

19. T. Utaka, T. Okanishi, T. Takeguchi, R. Kikuchi, K. Eguchi, Water gas shift reaction of reformed fuel over supported Ru catalysts, Appl. Catal. A: Gen. 245 (2003) 343–351.

20. T. Huang, T. Yu, S. Jhao, Weighting variation of water–gas shift in steam reforming of methane over supported Ni and Ni–Cu catalysts, Ind. Eng. Chem. Res. 45 (2006) 150–156.

21. A.V. Pattekar, M.V. Kothare, A microreactor for hydrogen production in microfuel cell applications, J. Microelectromech. Syst. 13 (2004) 7–18.

22. Y. Kawamura, N. Ogura, T. Yamamoto, A. Igarashi, A miniaturized methanol reformer with Si-based microreactor for a small PEMFC, Chem. Eng. Sci. 61 (2006) 1092–1101.

23. J. Bravo, A. Karim, T. Conant, G.P. Lopez, A. Datye, Wall coating of a CuO/ZnO/ Al2O3 methanol steam reforming catalyst for micro-channel reformers, Chem. Eng. J. 101 (2004) 113–121.

24. H. Yu, H. Chem, M. Pan, Y. Tang, K. Zeng, F. Peng, H. Wang, Effect of the metal foam materials on the performance of

methanol steam micro-reformer for fuel cells, Appl. Catal. A: Gen. 327 (2007) 106–113.

25. T. Terazaki, M. Nomura, K. Takeyama, O. Nakamura, T. Yamamoto, Development of multi-layered microreactor with methanol reformer for small PEMFC, J. Power Sources 145 (2005) 691–696.

26. W. Cai, F. Wang, A.V. Veen, C. Descorme, Y. Schuurman, W. Shen, C. Mirodatos, Hydrogen production from ethanol steam reforming in a micro-channel reactor, Int. J. Hydrogen Energy 35 (2010) 1152–1159.

27. A. Casanovas, M. Saint-Gerons, F. Griffon, J. Llorca, Autothermal generation of hydrogen from ethanol in a microreactor, Int. J. Hydrogen Energy 33 (2008) 1827–1833.

28. O. Görke, P. Pfeifer, K. Schubert, Kinetic study of ethanol reforming in a microreactor, Appl. Catal. A: Gen. 360 (2009) 232–241.

29. A. Casanovas, M. Domínguez, C. Ledesma, E. López, J. Llorca, Catalytic walls and micro-devices for generating hydrogen by low temperature steam reforming of ethanol, Catal. Today 143 (2009) 32–37.

30. A.Y. Tonkovich, S. Perry, Y. Wang, D. Qiu, T. LaPlante, W.A. Rogers, Microchannel process technology for compact methane steam reforming, Chem. Eng. Sci. 59 (2004) 4819–4824.

31. M. Levent, D.J. Gunn, M.A. El-Bousiffi, Production of hydrogen-rich gases from steam reforming of methane in an automatic catalytic microreactor, Int. J. Hydrogen Energy 28 (2003) 945–959.

32. Y. Seo, D. Seo, Y. Seo, W. Yoon, Investigation of the characteristics of a compact steam reformer integrated with a water–gas shift reactor, J. Power Sources 161 (2006) 1208–1216.

33. K.L. Yeung, S.M. Kwan, W.N. Lau, Zeolites in microsystems for chemical synthesis and energy generation, Top. Catal. 52 (2009) 101–110.

34. J.L.H. Chau, Y.S.S. Wan, A. Gavriilidis, K.L. Yeung, Incorporating zeolites in microchemical systems, Chem. Eng. J. 88 (2002) 187–200.

35. O.J. Kwon, S.-M. Hwang, J. Ahn, J.J. Kim, Silicon-based miniaturized-reformer for portable fuel cell applications, J. Power Sources 156 (2006) 253–259.

36. O.J. Kwon, S.-M. Hwang, I.K. Song, H. Lee, J.J. Kim, A silicon-based miniaturized reformer for high power electric devices, Chem. Eng. J. 133 (2007) 157–163.

37. O.J. Kwon, S.-M. Hwang, J.H. Chae, M.S. Kang, J.J. Kim, Performance of a miniaturized silicon reformer-PrOx-fuel cell system, J. Power Sources 165 (2007) 342–346.

38. S.H. Ahn, O.J. Kwon, I. Choi, J.J. Kim, Synergetic effect of combined use of Cu– ZnO–Al2O3 and Pt–Al2O3 for the steam reforming of methanol, Catal. Commun. 10 (2009) 2018–2022.

39. S.-M. Hwang, O.J. Kwon, S.H. Ahn, J.J. Kim, Silicon-based micro-reactor for preferential CO oxidation, Chem. Eng. J. 146 (2009) 105–111.

40. G.J. Seo, M.H. Youn, D.R. Park, J.C. Jung, I.K. Song, Hydrogen production by steam reforming of liquefied natural gas over mesoporous Ni–Al2O3 composite catalyst prepared by a single-step non-ionic surfactant-templating method, Catal. Lett. 132 (2009) 395–401.

41. A. Stefanescu, A.C. Veen, C. Mirodatos, J.C. Beziat, E. Duval-Brunel, Wall coating optimization for microchannel reactors, Catal. Today 125 (2007) 16–23.

42. J. Larminie, A. Dicks, Fuel Cell System Explained, John Wiley & Sons, New York, 2000.

43. M. Jong, A.H.M.E. Reinders, J.B.W. Kok, G. Westendorp, Optimizing a steammethane reformer for hydrogen production, Int. J. Hydrogen Energy 34 (2009) 285–292.

44. J.R. Rostrup-Nielsen, Coking on nickel catalysts for steam reforming of hydrocarbons, J. Catal. 33 (1974) 184–201.

45. D.L. Trimm, Coke formation and minimization during steam reforming reactions, Catal. Today 37 (1997) 233–238.

46. V.R. Choudhary, B.S. Uphade, A.S. Mamman, Simultaneous steam and CO2 reforming of methane to syngas over NiO/MgO/SA-5205 in presence and absence of oxygen, Appl. Catal. A-Gen. 168 (1998) 33–46.

47. N. Laosiripojana, W. Sangtongkitcharoen, S. Assabumrungrat, Catalytic steam reforming of ethane and propane over CeO2-doped Ni/Al2O3 at SOFC temperature: improvement of resistance toward carbon formation by the redox property of doping CeO2, Fuel 85 (2006) 323–332.

48. K.B. Lee, M.G. Beaver, H.S. Caram, S. Sircar, Novel thermal-swing sorptionenhanced reaction process concept for hydrogen production by lowtemperature steam-methane reforming, Ind. Eng. Chem. Res. 46 (2007) 5003–5014.

49. P. Panagiotopoulou, D.I. Kondarides, Effect of morphological characteristics of TiO2-supported noble metal catalysts on their activity for the water–gas shift reaction, J. Catal. 225 (2004) 327–336.

50. M.A. Edwards, D.M. Whittle, C. Rhodes, A.M. Ward, D. Rohan, M.D. Shannon, G.J. Hutchings, C.J. Kiely, Microstructural studies of the copper promoted iron oxide/chromia water–gas shift catalyst, Phys. Chem. Chem. Phys. 4 (2002) 3902–3908.

51. J.L. Ayastuy, M.A. Gutiérrez-Ortiz, J.A. González-Marcos, A. Aranzabel, J.R. González-Velasco, Kinetic of the low-temperature WGS reaction over a CuO/ZnO/Al2O3 catalyst, Ind. Eng. Chem. Res. 44 (2005) 41–50.

52. L. Qingfeng, H.A. Hjuler, N.J. Bjerrum, Phosphoric acid doped polybenzimidazole membrane: physiochemical characterization and fuel cell application, J. Appl. Electrochem. 31 (2001) 773–779.

Exergy Analysis and Optimization of a Hydrogen Production Process by a Solar-liquefied Natural Gas Hybrid Driven Transcritical CO2 Power Cycle

Zhixin Sun[a], Jiangfeng Wang[a], Yiping Dai[a], and
Jihong Wang[b]

[a]School of Energy and Power Engineering, Xi'an Jiaotong University,
No. 28 Xianning West Road, Xi'an 710049, PR China
[b]School of Engineering, University of Warwick, Coventry, CV4 7AL,
UK

ABSTRACT

A solar transcritical CO_2 power cycle for hydrogen production is studied in this paper. Liquefied Natural Gas (LNG) is utilized to condense the CO_2. An exergy analysis of the whole process is performed to evaluate the effects of the key parameters, including the boiler inlet temperature, the turbine inlet temperature, the turbine inlet pressure and the condensation temperature, on the system power outputs and to guide the exergy efficiency improvement. In addition, parameter optimization is conducted via Particle Swarm Optimization to maximize the exergy efficiency of hydrogen production. The exergy analysis indicates that both the solar and LNG equally provide exergy to the CO_2 power system. The largest amount of exergy losses occurs in the solar collector and the condenser due to the great temperature differences during the heat transfer process. The exergy loss in condenser could be greatly reduced by increasing the LNG temperature at the inlet of the condenser. There exists an optimum turbine inlet pressure for achieving the maximum exergy efficiency. With the optimized turbine inlet pressure and other parameters, the system is able to provide 11.52 kW of cold exergy and 2.1 L/s of hydrogen. And the exergy efficiency of hydrogen production could reach 12.38%.

INTRODUCTION

It is well known that using fossil fuels as the primary energy source has brought serious impact onto the environment and led to energy crisis. In order to mitigate the environmental impact, great efforts have been devoted to develop clean, renewable energy sources. Solar energy, as one of the most promising renewable energy resources, has gained rapid development.

Except for the photovoltaic generation, Rankine power cycle is the alternative most competitive approach to produce electricity via solar energy. In order to make better use of the thermal energy, organic substances and carbon dioxide have been utilized in power cycle instead of water. A lot of studies have been carried out on selecting the working fluids [1], [2] and [3]. Among all these working fluids, carbon dioxide has drawn more and more attention recently since it is non-explosive, non-flammable, nearly non-toxic and is naturally abundant [3]. In addition, CO_2 has favorable thermodynamic properties [4]. It is easy for CO_2 to reach its supercritical state (7.38 MPa and 31.1 °C). The temperature glide of supercritical CO_2 offers a better temperature profile match to the heat source temperature than subcritical fluids since there is no isothermal evaporation process for supercritical fluids. A number of references can be found about the solar-driven supercritical CO_2 system [5], [6] and [7].

There are two main issues associated with the solar-driven supercritical CO_2 system. One is the time-varying solar radiation, and the other is the condensation of CO_2. The difficulties of the former issue can be alleviated by the using of thermal storage tank or the solar-wind hybrid system [8], [9] and [10]. Moreover, intermittent hydrogen production has been identified as the most promising way to solve the problem [11], [12] and [13]. Hydrogen is clean. When it is converted into electricity via a fuel cell, the only by-products are harmless water and heat. Moreover, hydrogen contains more energy per unit mass than other fuels and it can be produced from numerous sources, such as methane, coal, biomass, water, etc. Nowadays, hydrogen has been considered as an ideal energy carrier in the foreseeable future. Many countries in the world have realized the vision of hydrogen economy. Hydrogen production technology, especially renewable hydrogen production, is developing rapidly.

A standard Rankine cycle comprises a condensation process. Exhaust vapor from the turbine is condensed to liquid and then pumped to the boiler to continue the cycle. As the critical temperature of CO_2(31.1 °C) is very close to the environmental temperature, even lower than that in summer, the condensation difficulty becomes the main factor that restrict the development of supercritical CO_2 cycle. Many studies use cooling water as the heat sink where the condensation of CO_2 takes place just below the critical temperature [14] and [15]. With the development of cryogenic exergy recovery of LNG, LNG has been introduced as the heat sink to cool the CO_2 in the thermal cycles. Zhang and Lior [16] proposed a LNG fueled quasi-combined system of Rankine-like cycle and Brayton cycle with CO_2 as the working fluid and LNG as the heat sink. Lin et al. [17] presented a transcritical CO_2 power cycle for heat recovery of a gas turbine exhaust with LNG as the heat sink. Song et al. [7] firstly employed LNG as the heat sink in their solar-driven transcritical CO_2 cycle. And parameter analysis was conducted based on the first law of thermodynamics. However, thermodynamic analysis based on the first law, known as energy analysis, only reflects the conservation of energy. In the energy analysis of a solar-LNG transcritical CO_2 cycle, the solar collectors provide thermal heat to the power cycle. One portion of the heat is converted to mechanical power, and the other portion is taken away by the LNG. The LNG acts as a heat sink. Whereas from the viewpoint of the second thermodynamics law, the LNG provides cryogenic exergy to the power cycle rather than taking away exhaust heat. The LNG acts as an exergy provider as the solar collectors do. Hence, exergy analysis is quite essential for the solar-LNG hybrid driven transcritical CO_2 cycle.

In this paper, exergy analysis based on the second law of thermodynamics was conducted to evaluate the solar-LNG hybrid driven transcritical CO_2 cycle. A thermal storage tank and an electrolyzer for hydrogen production were added to alleviate the influence of intermittent solar radiation. Parameter optimization was also conducted via particle swarm optimization (PSO) to obtain the maximum exergy efficiency.

The organization of the paper is as follows. Section 2 deals with the system description. In Section 3, we present the model for exergy analysis. Section 4 introduces the optimization algorithm and Section 5 outlines the results of exergy analysis and optimization. Conclusions appear in Section 6.

SYSTEM DESCRIPTION

The schematic diagram of the solar-LNG hybrid driven transcritical CO_2 hydrogen production system is shown in Fig. 1. This system is composed of three subsystems: the solar collector subsystem, the transcritical CO_2 power subsystem, and the hydrogen production subsystem.

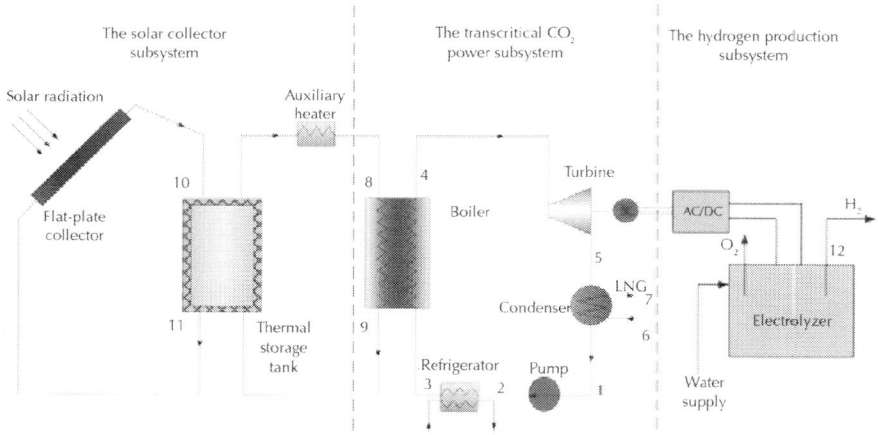

Figure 1: Schematic diagram of the transcritical CO_2 hydrogen production system.

The solar collector subsystem comprises three key components: a set of solar collectors, a thermal storage tank and an auxiliary heater. The liquid flat-plate collector is employed as the source of energy in the system due to its low cost and wide applications. The thermal storage tank is installed to store the thermal energy produced from solar energy that provides heat to the downstream power cycle when there is no or little solar radiation. The solar collectors absorb the solar radiation to heat the water. Then the hot water enters the storage tank and transfers its thermal energy to the water in the tank through mixing. Two streams of water flow out of the storage tank. One goes back into the collectors to continue the heat absorption cycle. The other flows into the boiler to heat the

CO_2 and then returns to the tank. An auxiliary heater is added as the backup heat source. The mathematical descriptions of the solar collectors and the thermal storage tank can refer to [7].

The transcritical CO_2 power subsystem is made up of five components: a boiler, a turbine, a condenser, a pump and a refrigerator. Fig. 2 illustrates the corresponding cycle composed of the following processes:

1–2: a non-isentropic compression process in the pump;

2–3: a non-isobaric heat absorption process in the refrigerator;

3–4: a non-isobaric heat absorption process in the boiler;

4–5: a non-isentropic expansion process in the turbine;

5–1: an isobaric heat rejection process in the condenser.

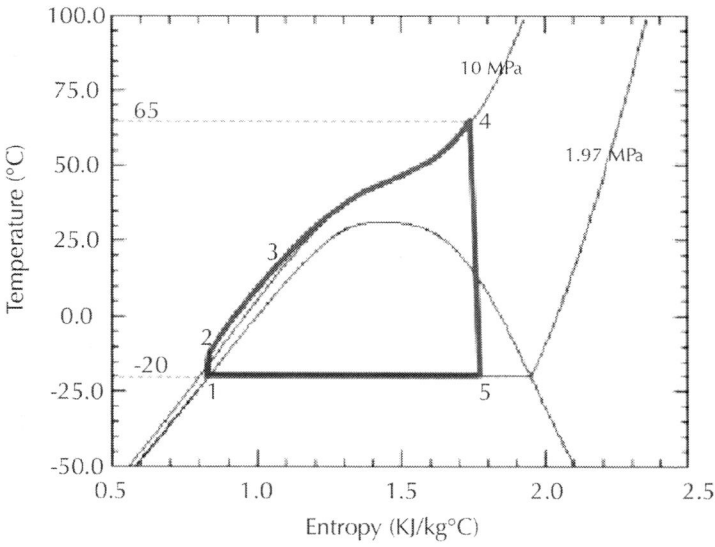

Figure 2: T–s diagram for the transcritical CO_2 power cycle.

The supercritical CO_2 from the boiler enters the steam turbine and expands to achieve a lower pressure which drives a generator to produce electricity. The exhaust vapor of the turbine is condensed to liquid in the condenser and LNG is chosen as the heat sink. The

CO_2 is subsequently pumped to a supercritical pressure. Due to the low temperature in the condenser, the CO_2 temperature at the outlet of the pump is lower than the environmental temperature. Hence, we proposed to add a heat exchanger, acting as a refrigerator, to recover the cold energy in CO_2 as well as to save the heat energy provided by the thermal storage tank. The CO_2 is nearly heated to the environmental temperature in the refrigerator and then flows into the boiler. In the boiler, the carbon dioxide is lifted to a higher temperature by the energy from the thermal storage tank.

To simplify the system, the following assumptions are made:

- The system reaches a steady state, the kinetic and potential energies of fluids are neglected.
- The condenser outlet state is saturated liquid.
- The pressure drops of CO_2 in the refrigerator and boiler are both assumed to be 2%.
- The LNG is vaporized at a constant pressure of 0.6MPa, which equals to the liquefied pressure in a LNG filling station.

Water electrolysis is one of the most basic and widely used technologies in hydrogen production, and especially for hydrogen production from renewable sources [12]. The alternating current produced by the generator is firstly converted into direct current, and then used to drive the electrolyzers for hydrogen production.

The voltage corresponding to Higher Heating Value (HHV) is 1.48 V. This represents the thermoneutral voltage where hydrogen and oxygen would be produced with 100% thermal efficiency (i.e., no waste heat produced from the reaction). This is determined by dividing the HHV (285,840 J/mol) of H_2 by the Faraday constant (96,485 coulombs/mole) and the number of electrons needed to create a molecule of H_2. The voltage corresponding to the Lower Heating Value (LHV, 237,122 J/mol) is 1.23 V [18]. In this study, all the calculations are based on the HHV rather than LHV since HHV accounts for the total amount of energy in the electrolysis process.

If one mole of hydrogen is burned, it would produce one mole of water and discharge the heat amount of 285,840 J. So when one mole of water is split into one mole of hydrogen and half mole of oxygen, the full 285,840 J of energy is required. That is, for an electrolyzer with voltage of 1.48 V and efficiency of 100%, 285,840 J of electricity would generate one mole of hydrogen, equating to 141.8 MJ of electricity for one kilogram of hydrogen through unit conversion. With consideration of less than perfect efficient systems, more electrical energy will be required.

There are three types of water electrolyzers available in industry: (1) alkaline electrolyzer with lowest cost and efficiency of about 80%, (2) proton exchange membrane electrolyzer with highest cost and efficiency of about 94.4%, and (3) solid oxide electrolyzer with median cost and efficiency of about 90%[19]. An alkaline electrolyzer is chosen in this study and a total efficiency of 77% is assumed considering the energy dissipations of AC/DC converter and other equipment.

EXERGY ANALYSIS OF THE SYSTEM

Exergy is the maximum amount of work that a system can generate when it comes to equilibrium with a reference environment. Exergy analysis is usually employed to specify the exergy losses in devices and identify the potential ways for exergy saving in each process.

The exergy at a state point i can be expressed as

$$E_i = m[(h_i - h_0) - T_0(s_i - s_0)] \tag{1}$$

Exergy analysis of a complex system can be achieved by analyzing the exergy losses in each component of the system separately. The exergy loss in each component equals the difference between the exergy input and the output

$$I = \sum E_{in} - \sum E_{out} \tag{2}$$

The exergy loss in the boiler is

$$I_B = E_8 + E_3 - E_9 - E_4 \tag{3}$$

The exergy loss in the turbine can be calculated by

$$I_{TUR} = E_4 - W_{TUR} - E_5 \tag{4}$$

The exergy loss in the condenser is obtained from

$$ICON = E_5 + E_6 - E_1 - E_7 \tag{5}$$

The exergy loss in the pump is

$$IPUMP = WPUMP + E_1 - E_2 \tag{6}$$

The exergy loss in the refrigerator is

$$IREF = E_2 - E_3 \tag{7}$$

The exergy loss in the hydrogen production devices is

$$IH_2 = WTUR - WPUMP - E_{12} \tag{8}$$

where E_{12} denotes the exergy of H_2 and approximately equates to the HHV of H_2.

The solar collector exergy efficiency CLT is defined as the exergy increase of the water divided by the solar radiation exergy E_S absorbed by the collector.

$$\eta_{CLT} = \frac{E_{10} - E_{11}}{E_S} \tag{9}$$

$$E_S = P_S \left[1 + \frac{1}{3}\left(\frac{T_0}{T_S}\right)^4 - \frac{4}{3}\frac{T_0}{T_S} \right] \tag{10}$$

where P_S is the heat absorbed by the collector and T_S is the solar radiation temperature of 6000 K [20].

The instantaneous exergy efficiency of the whole solar-LNG hybrid driven hydrogen production system S_H_2 can be given as

$$\eta_{S-H_2} = \frac{E_{12}}{E_S + (E_6 - E_7)} \tag{11}$$

The heat that the collector absorbed does not balance with the heat that the CO_2 cycle consumed because of the existence of the thermal storage tank. Hence, the hydrogen production exergy efficiency with regard to the CO_2 cycle $CO_2 - H_2$ is given as

$$\eta_{CO_2 - H_2} = \frac{E_{12}}{(E_8 - E_9) + (E_6 - E_7)} \tag{12}$$

PARTICLE SWARM OPTIMIZATION (PSO)

As it is quite difficult or impossible for the traditional optimization algorithms to solve the optimization problems of a complex system, many modern optimization algorithms have been proposed. Particle swarm optimization (PSO), which was developed with the inspiration from the intelligence in movement of a bird flock or fish school, is one of the relatively new evolutionary computational methods in modern optimization technologies. It has simple structure and its optimization process shows a clear physical meaning. Genetic algorithm is also one of the most widely used modern optimization algorithms. The comparisons between PSO, Genetic Algorithm and traditional optimization techniques are listed in Table 1.

Table 1: Comparison between PSO and other optimization techniques

Term	PSO	Genetic algorithm	Traditional optimization algorithm
Complex problems	Easy	Easy	Difficult
Computation time	Long	Long	Short
Extensibility	Easy	Easy	Difficult
Easy to implement	Yes	No	Yes
Few parameters to adjust	Yes	No	Yes

PSO is a population-based algorithm proposed by Kennedy and Eberhart [21] and has been widely and successfully applied to solve complex optimization problems [22], [23] and [24]. The PSO algorithm is based on a wide search population (called a swarm) that consists of many individuals (called particles). When the swarm is searching for target solution as a collective activity, each particle in PSO flies through the search space and adjusts its velocity with time according to the successful "flying" or "swimming" experience of its own and the whole swarm. Hence, the swarm is capable of moving toward the best solutions.

The flow chart for PSO implementation is given in Fig. 3 and the mathematical description of PSO is as follows. Suppose that the search space is m-dimensional, then the current position and velocity of the i th particle can be represented by $Xi = T[X_{i1}, X_{i2}, ..., X_{im}]$ and $Vi = T[V_{i1}, V_{i2}, ..., V_{im}]$ respectively, where $i = 1,2, ..., N$ and N is the number of particles in the swarm.

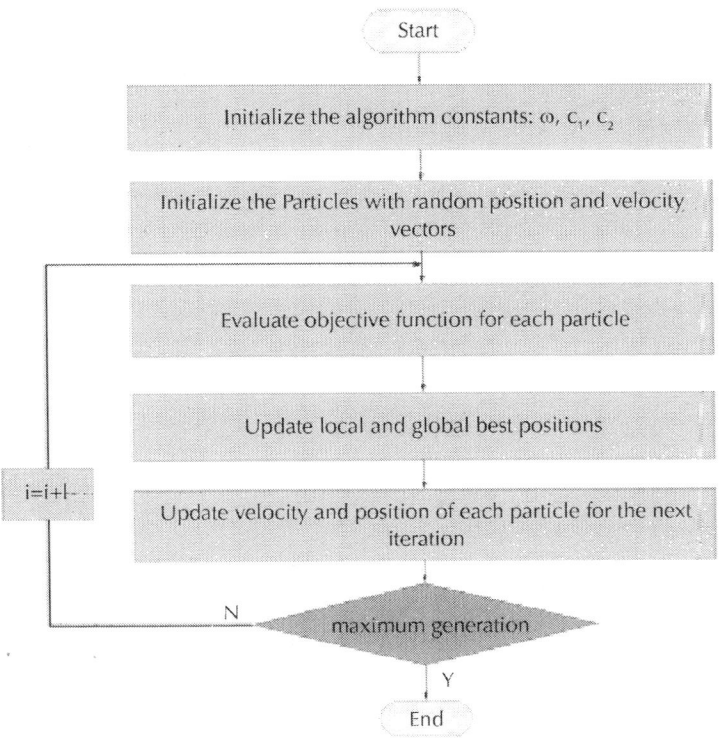

Figure 3: Flow chart for PSO implementation.

Particle i can remember the best position it ever visited, which is known as the local best position $Li = T[L_{i1}, L_{i2}, ..., L_{im}]$. It can also get the best position that the whole swarm found, known as the global best position $G = T[G_1, G_2, ..., Gm]$. The first position and velocity of Particle i are randomly initialized. Afterwards, Particle i adjusts its velocity of iteration $t + 1$ according to the local and global best positions, as well as the velocity and position of iteration t, by

$$Vi(t+1)=\omega Vi(t)+c_1 R_1(Li(t)-Xi(t))+c_2 R_2(G(t)-Xi(t))(13)$$

where   is the inertia coefficient which is employed to manipulate the impact of the previous history of velocities on the current velocity. c_1 and c_2 are the accelerate coefficients towards local and global best positions. R_1 and R_2 are uniformly distributed random real numbers on the interval [0,1].

With the updated velocity, the position of Particle i in the iteration $t + 1$ can be obtained by

$$Xi(t+1)=Xi(t)+Vi(t+1) \qquad\qquad (14)$$

The updates of velocity and position continue until a specified number of iteration has been exceeded.

In this study, the objective function is designed for maximizing exergy efficiency of hydrogen production.

SIMULATION RESULTS AND DISCUSSION

The results for simulation study of the transcritical CO_2 power cycle are presented in this section. Xi'an, China (34.27N, 108.95E) is selected as the example location and June 21st is chosen as the typical day to conduct the simulation. The flat-plate collector, pointing to due south, is tilted at an angle of 8° with the horizontal to receive the maximum local radiation [7]. The number of solar collectors is presumed to be 75, with a total area of 450 m², which can be adjusted in accordance with the load demanded. The simulation code is written in Fortran language. The thermodynamic properties of the working fluids are calculated by REFPROP 6.01 which is developed by the National Institute of Standards and Technology of the United States.

Exergy Analysis

Fig. 4 shows the variations of ambient temperature and the water temperature at the outlet of the thermal storage tank, along with the collector heat gain and power consumption of the CO_2 cycle on a daily basis. It can be seen that from 7:00 to 8:00, the energy consumption exceeds the heat input, resulting in a decrease in water temperature. Thereafter the collector heat gain surpasses the energy consumption until 15:00. Consequently, the tank water temperature begins to increase steadily from 8:00 and culminates at around 15:00. Then the water temperature at the outlet of thermal storage tank drops as the useful heat gain decreases. Since the tank water directly provides the heat for the CO_2 power cycle, the power output undergoes almost the same variation trends with the water temperature. Hence, the water storage tank behaves as a buffer to provide stable and continuous heat for the power generation. The tank temperature fluctuates by about 5 °C. The magnitude of this fluctuation primarily depends on the tank capacity. Large capacity would alleviate the temperature fluctuation.

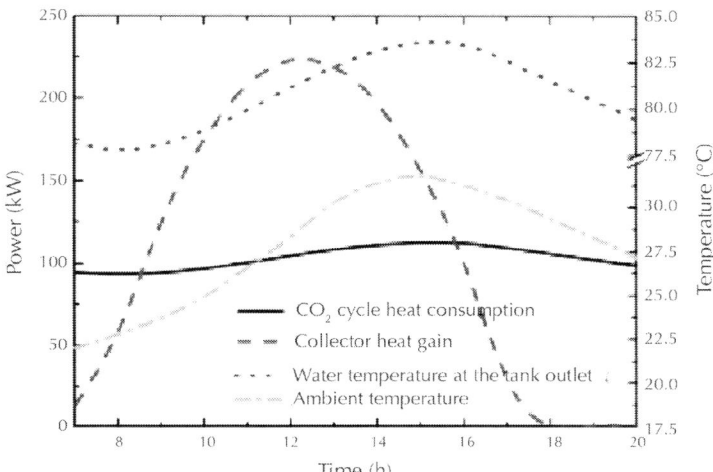

Figure 4: Main parameter variations of the solar collector subsystem over a day.

The simulation conditions of the system are displayed in Table 2. Despite the ambient temperature varies in a day, a constant temperature of 25 °C with the atmospheric pressure of 0.1 MPa is set as the reference state for exergy calculation. Since the operation condition varies with time, a typical time of 15:00, when the system is operating near its peak load, is chosen to perform the simulation. Table 3 shows the thermodynamic parameters at each node of the power system and Table 4 shows the performance of the system. It is found that at 15:00, the system is capable of generating about 1.12 L of H_2 per second with a system instantaneous exergy efficiency of 4.98%.

Table 2: Simulation conditions of the transcritical CO_2 cycle for hydrogen production

Term	Value
Ambient temperature (°C)	25
Ambient pressure (MPa)	0.1
Molar composition of ambient air	N_2 = 75.60%; O_2 = 20.34%;
	H_2O = 3.12%; CO_2 = 0.03%
	Other gases = 0.91%
Time	15:00
Collector tilt angle	8°
Inner diameter of the absorber tube (m)	0.014
Outer diameter of the absorber tube (m)	0.018
Total surface area of absorber plates (m²)	450
Volume of the storage tank (m³)	86
Boiler inlet temperature (°C)	20
Turbine inlet temperature (°C)	65
Turbine inlet pressure (MPa)	10
Condensation temperature (°C)	−20
LNG inlet temperature (°C)	−161.48
LNG pressure (MPa)	0.6
Turbine isentropic efficiency	80%
Pump isentropic efficiency	80%

Table 3: Results of the transcritical CO_2 cycle for hydrogen PRODUCTION

State	t (°C)	p (MPa)	h (kJ/kg)	s (kJ/kg/K)	e (kJ/kg)	m (kg/s)
1	−20.00	1.97	154.45	0.83	217.02	0.558
2	−14.69	10.41	164.60	0.84	224.82	0.568
3	20.00	10.20	242.36	1.12	218.38	0.568
4	65.00	10.00	438.82	1.75	229.23	0.568
5	−20.00	1.97	396.19	1.79	174.05	0.568
6	−161.48	0.60	0.733	−0.004	1083.99	0.176
7	−30.00	0.60	783.06	5.29	288.32	0.176
8	83.33	0.11	348.97	1.11	22.58	0.8
9	50.00	0.11	209.42	0.70	5.27	0.8
10	89.73	0.11	375.86	1.19	25.61	5.799
11	83.33	0.11	348.97	1.11	22.58	5.799

Table 4: Performance of the transcritical CO_2 cycle for hydrogen production

Term	Value
Collector exergy gain (kW)	145.60
Turbine power (kW)	24.22
Pump power (kW)	5.77
Net power output (kW)	18.45
Refrigeration output (kW)	3.66
Hydrogen production rate (L/s)	1.12
Collector instantaneous exergy efficiency	12.10%
Exergy efficiency of H_2 to collector	4.98%
Exergy efficiency of H_2 to CO_2	9.26%

Exergy analysis has been performed in two cases. case 1 starts from the solar collector process to the H_2 process. As there is a buffer, the thermal storage tank, between the solar collector and the downstream processes, the heat that the collector absorbed does not balance with the heat that the CO_2 cycle consumed. Hence, the solar collector is excluded in the analysis of Case 2. The exergy losses of each process are shown in Table 5. It is found in Case 1 that the solar collector and the LNG contribute nearly the same amount of exergy to the system. And they also account for the largest exergy losses. These two amounts of exergy losses result from the great temperature differences between the solar and the hot water in the collector and the temperature differences between the CO_2 and the LNG in the condenser. The other exergy losses are related to the irreversibilities of the rest processes.

Table 5: The exergy inputs, outputs, and losses of each component in the system

Case Value (kW)		1. Solar collector to H$_2$		2. CO$_2$ cycle to H$_2$	
		Percentage	Value (kW)	Percentage	
Exergy inputs	Collector	145.60	51.04	/	/
	Boiler	/	/	13.85	9.02
	Condenser	139.69	48.96	139.69	90.98
Exergy out-puts	H$_2$	14.21	4.98	14.21	9.26
	Refrigeration	3.66	1.28	3.66	2.38
Exergy losses	Collector	127.98	44.86	/	/
	Storage tank	3.77	1.32	/	/
	Boiler	7.69	2.70	7.69	5.01
	Turbine	7.13	2.50	7.13	4.64
	Condenser	115.27	40.40	115.27	75.07
	Pump	1.33	0.47	1.33	0.87
	H$_2$ production	4.25	1.49	4.25	2.77

In the analysis of Case 2, the LNG provides approximately 91% of exergy since most of the solar exergy is lost in the collector. The exergy loss in the condenser is about 75%. The approach to reduce the exergy loss in the condenser is to increase the LNG inlet temperature. Through the combination with the cascade utilization of LNG cold energy [25] and [26], the low temperature LNG could be firstly preheated in other processes, such as dry ice production, then employed to condense the exhaust CO_2. If the inlet LNG (or rather natural gas) temperature increases to −70 °C, the exergy loss in condenser would drop from 115.27 to 22.25 kW, and the whole system efficiency would rises from 4.98% to 7.39%.

The parametric analysis is performed to evaluate the effects of some key parameters, namely, the boiler inlet temperature, the turbine inlet temperature, the turbine inlet pressure and the condensation temperature, on the system power and efficiency. When one parameter varies, the other parameters are kept constant as shown in Table 2. Fig. 4, Fig. 5, Fig. 6 and Fig. 7 present the relationship between these key parameters and the system performance.

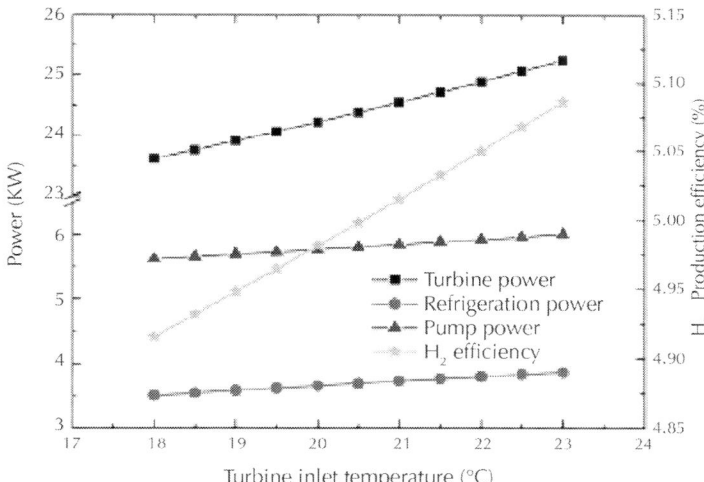

Figure 5: Effect of the boiler inlet temperature on the system power and efficiency.

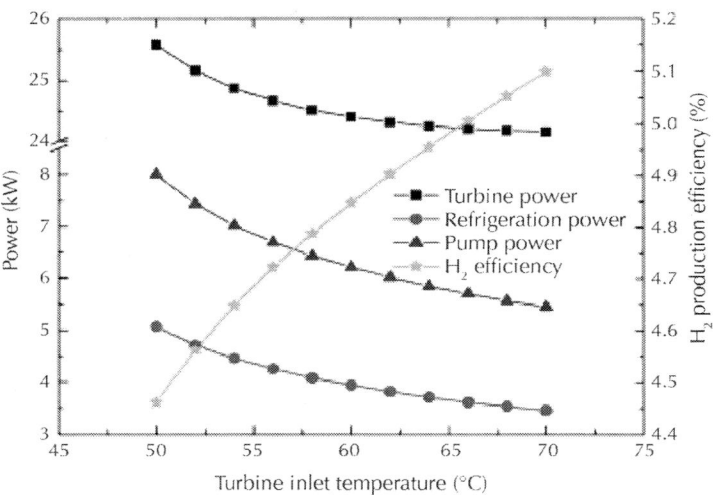

Figure 6: Effect of the turbine inlet temperature on the SYSTEM power and efficiency.

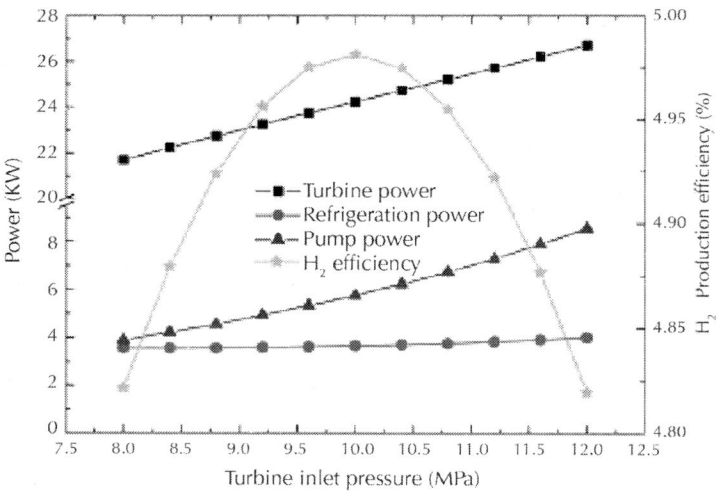

Figure 7: Effect of the turbine inlet pressure on the system power and efficiency.

Fig. 5 shows the effect of the boiler inlet temperature on the system power and efficiency. As the boiler inlet temperature increases, the CO_2 absorbs more heat from the environment other than from the boiler. Hence, the boiler could afford more CO_2, which means the flow rate of CO_2 rises. With the increase of the flow rate of CO_2, the turbine output, the refrigeration power and the H_2 production output all rise. Thus, more effort should be devoted to reduce the difference between the boiler inlet temperature and environment temperature.

As shown in Fig. 6, both the turbine power and the pump power decrease as the turbine inlet temperature increases. Since the drop in pump power is greater than that of the turbine power, the net power output for H_2 production increases. The refrigeration power decreases by about 1 kW with a 10 °C increase in the turbine inlet temperature.

It can be observed from Fig. 7 that as the turbine inlet pressure increases, the H_2 production efficiency increases firstly and then declines. The increase in the efficiency can be explained by the considerable increase in the turbine power output. Afterward, the increase in the turbine power is overtaken by the increase in the power consumption of the pump. Thus, the net power output left for H_2production decreases after it reaches the peak. The turbine inlet pressure has little effect on the refrigeration output.

It is shown in Fig. 8 that the condensation temperature has considerable influences on the system efficiency. As the condensation temperature decreases, the enthalpy difference between CO_2 at the inlet and outlet of the turbine rises. Despite the fact that the flow rate of CO_2 decreases, the H_2production efficiency increases by about 1% with a 10 °C drop in the condensation temperature. And the refrigeration power increases by about 2 kW with a 10 °C drop in the condensation temperature.

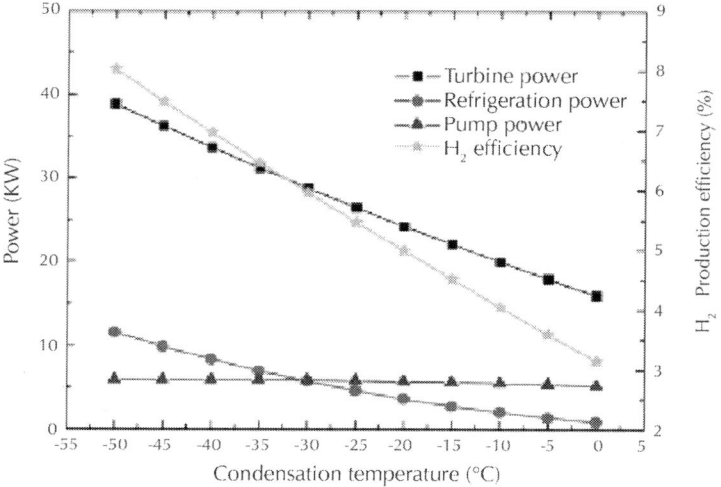

Figure 8: Effect of the condensation temperature on the system power and efficiency.

Optimization Results

As stated above, some parameters have great influence on the system exergy efficiency. Hence, parameter optimization is conducted via PSO algorithm to obtain the optimal design parameters of the system. The optimization objective is to reach the maximum exergy efficiency of hydrogen production. The boiler inlet temperature, the condensation temperature, the turbine inlet temperature and pressure, as well as the LNG inlet temperature, are the optimization parameters. Table 6 presents the condition of the parameter optimization and the optimization results are listed in Table 7.

Table 6: The condition of the parameter optimization

Number of particles	20
Inertia coefficient, ω	0.6
Accelerate coefficient of local best, c_1	2

Accelerate coefficient of global best, c_2	2
Maximum generation	500
The range of boiler inlet temperature (°C)	18–23
The range of turbine inlet temperature (°C)	50–75
The range of turbine inlet pressure (MPa)	8–14
The range of condenser temperature (°C)	−50 to 0
The range of LNG inlet temperature (°C)	−161.48 to −70

Table 7: The optimization results of the SYSTEM

Boiler inlet temperature (°C)	23
Turbine inlet temperature (°C)	75
Turbine inlet pressure (MPa)	11.67
Condenser temperature (°C)	−50
LNG inlet temperature (°C)	−70
Turbine work(kW)	41.66
Pump work(kW)	7.16
Net power output(kW)	34.50
Refrigeration output (kW)	11.52
Hydrogen production rate (L/s)	2.10
Exergy efficiency of collector to electricity η_{s-h2}	12.38%
Exergy efficiency of CO_2 to H_2 η_{co2-h2}	32.05%

Compared with the results in Table 4, the system efficiency rises from 4.98% up to 12.38% due to the increases in boiler inlet temperature, turbine inlet temperature, turbine inlet pressure and the LNG inlet temperature as well as the decreases in condensation temperature.

CONCLUSIONS

In the present study, a solar-LNG hybrid driven transcritical CO_2 power cycle for hydrogen production is investigated based on the exergy analysis. The influences of some key parameters on the system performance are examined, and the optimal values of these parameters are obtained via particle swarm optimization. The main conclusions drawn from the study are listed as follows:

- Both the solar collector and the LNG equally provide exergy to the hydrogen production system.
- The exergy loss in condenser could be greatly reduced by increasing the LNG inlet temperature.
- The system exergy efficiency increases as the boiler inlet temperature and the turbine inlet temperature rises and as the condensation temperature decreases. There exists an optimum turbine inlet pressure for maximum exergy efficiency.
- With the optimized parameters, the system is able to provide 11.52 kW of cold exergy and 2.1 L/s of hydrogen. And the hydrogen production efficiency of 12.38% could be reached.

As the intensity of solar radiation varies with time of day and year, static model cannot evaluate the dynamic performance over a day or a year. Hence, dynamic model will be established to study the system behavior over a period of time.

ACKNOWLEDGMENTS

The authors gratefully acknowledge the support of the national Key Technology Research and development Program (No. 2011BAA05B03), the National High Technology Research and Development Program (No. SS2012AA053002) of China and the funding support from the University of Warwick Strategic Awards.

REFERENCES

1.   Delgado-Torres AM, Garci´a-Rodri´guez L. Analysis and optimization of the low-temperature solar Organic Rankine Cycle (ORC). Energy Conversion and Management 2010; 51(12):2846e56.

2.   Liu BT, Chien KH, Wang C-C. Effect of working fluids on organic rankine cycle for waste heat recovery. Energy 2004; 29(8):1207e17.

3.   Chen Y, Lundqvist P, Johansson a, Platell P. A comparative study of the carbon dioxide transcritical power cycle compared with an organic rankine cycle with R123 as working fluid in waste heat recovery. Applied Thermal Engineering 2006;26(17):2142e7.

4.   Beckman EJ. Supercritical and near-critical CO_2 in green chemical synthesis and processing. Journal of Supercritical Fluids 2004;28(2):121e91.

5.   Zhang XR, Yamaguchi H. An experimental study on evacuated tube solar collector using supercritical CO_2. Applied Thermal Engineering 2008;28(10):1225e33.

6.   Zhang XR, Yamaguchi H, Uneno D, Fujima K, Enomoto M, Sawada N. Analysis of a novel solar energy-powered rankine cycle for combined power and heat generation using super critical carbon dioxide. Renewable Energy 2006;31(12):1839e54.

7.   Song Y, Wang J, Dai Y, Zhou E. Thermodynamic analysis of a transcritical CO_2 power cycle driven by solar energy with liquefied natural gas as its heat sink. Applied Energy 2012;92: 194e203.

8.   Badescu V. Optimal control of flow in solar collector systems with fully mixed water storage tanks. Energy Conversion and Management 2008;49(2):169e84.

9.   Eldighidy SM, Taha IS. Water in a flat plate solar tank and an organic rankine cycle. Solar Energy 1983;31(5):455e61.

10. Reichling J, Kulacki F. Utility scale hybrid windesolar thermal electrical generation: a case study for Minnesota. Energy 2008;33(4):626e38.

11. Zhang XR, Yamaguchi H, Cao Y. Hydrogen production from solar energy powered supercritical cycle using carbon dioxide. International Journal of Hydrogen Energy 2010; 35(10):4925e32.

12. Dincer I. Green methods for hydrogen production. International Journal of Hydrogen Energy 2012;37(2):1954e71.

13. Granovskii M, Dincer I, Rosen M a. Exergetic life cycle assessment of hydrogen production from renewables. Journal of Power Sources 2007;167(2):461e71.

14. Zhang XR, Yamaguchi H, Uneno D. Experimental study on the performance of solar rankine system using supercritical CO_2. Renewable Energ 2007;32(15):2617e28.

15. Wang J, Sun Z, Dai Y, Ma S. Parametric optimization design for supercritical CO_2 power cycle using genetic algorithm and artificial neural network. Applied Energy 2010;87(4):1317e24.

16. Zhang N, Lior N. A novel near-zero CO_2 emission thermal cycle with LNG cryogenic exergy utilization. Energy 2006; 31(10, 11):1666e79.

17. Lin W, Huang M, He H, Gu A. A transcritical CO_2 Rankine cycle with LNG cold energy utilization and liquefaction of CO_2 in gas turbine exhaust. Journal of Energy Resources e ASME 2009; 131(4):042201e5.

18. Gibson T, Kelly N. Optimization of solar powered hydrogen production using photovoltaic electrolysis devices. International Journal of Hydrogen Energy 2008;33(21): 5931e40.

19. Ni M, Leung M, Sumathy K, Leung D. Potential of renewable hydrogen production for energy supply in Hong Kong. International Journal of Hydrogen Energy 2006;31(10): 1401e12.

20. Hepbasli A. A key review on exergetic analysis and assessment of renewable energy resources for a sustainable

future. Renewable & Sustainable Energy Reviews 2008;12(3): 593e661.

21. Kennedy J, Eberhart R. Particle swarm optimization. Proceedings of ICNN'95-International Conference on Neural Networks 1995;4:1942e8.

22. Wu XJ, Huang Q, Zhu XJ. Thermal modeling of a solid oxide fuel cell and micro gas turbine hybrid power system based on modified LS-SVM. International Journal of Hydrogen Energy 2011;36(1):885e92.

23. Vachirasricirikul S, Ngamroo I, Kaitwanidvilai S. Application of electrolyzer system to enhance frequency stabilization effect of microturbine in a microgrid system. International Journal of Hydrogen Energy 2009;34(17):7131e42.

24. Zhong ZD, Zhu XJ, Cao GY, Shi JH. A hybrid multi-variable experimental model for a PEMFC. Journal of Power Sources 2007;164(2):746e51.

25. Lu T, Wang KS. Analysis and optimization of a cascading power cycle with liquefied natural gas (LNG) cold energy recovery. Applied Thermal Engineering 2009;29(8):1478e84.

26. Deng S, Jin H, Cai R, Lin R. Novel cogeneration power system with liquefied natural gas (LNG) cryogenic exergy utilization. Energy 2004;29(4):497e512.

Thermodynamic Design of a Cascade Refrigeration System of Liquefied Natural Gas by Applying Mixed Integer Non-linear Programming

Meysam Kamalinejad, Majid Amidpour, and S.M. Mousavi Naeynian

Department of Mechanical Engineering, K.N. Toosi University of Technology, Tehran 1999143344, Iran

ABSTRACT

Liquefied natural gas (LNG) is the most economical way of transporting natural gas (NG) over long distances. Liquefaction of

NG using vapor compression refrigeration system requires high operating and capital cost. Due to lack of systematic design methods for multistage refrigeration cycles, conventional approaches to determine optimal cycle are largely trial-and-error. In this paper a novel mixed integer non-linear programming (MINLP) model is introduced to select optimal synthesis of refrigeration systems to reduce both operating and capital costs of an LNG plant. Better conceptual understanding of design improvement is illustrated on composite curve (CC) and exergetic grand composite curve (EGCC) of pinch analysis diagrams. In this method a superstructure representation of complex refrigeration system is developed to select and optimize key decision variables in refrigeration cycles (*i.e.* partition temperature, compression configuration, refrigeration features, refrigerant flow rate and economic trade-off). Based on this method a program (LNG-Pro) is developed which integrates VBA, Refprop and Excel MINLP Solver to automate the methodology. Design procedure is applied on a sample LNG plant to illustrate advantages of using this method which shows a 3.3% reduction in total shaft work consumption.

INTRODUCTION

Natural gas (NG) is an attractive source of clean fossil fuel and the third primary energy source after crude oil and coal. It is also the fastest growing and second largest energy source for electricity generation. In 2012, NG consumption was 2987.1 million tons oil equivalent, or about 24% of the total primary energy consumed worldwide. World's primary energy consumption had an average growth rate of 2.6% during the last 10 years, but LNG consumption growth rate was 7.85% [1]. This growth means a promising future for LNG industry. Most NG reserves are offshore and away from demand centers. Liquefying NG and transporting it to distances further from 3000 km is the most economical way to export it to consuming market. LNG industry is very energy extensive and industrial size LNG plants consume around 1181 kJ of energy to liquefy 1 kg of NG [2]. Heat integration inside a cycle or between different cycles of a cascade can greatly reduce shaft work consumption. Therefore, energy is an immediate concern in LNG industry. Such refrigeration system involves some of the largest compressors in the world, usually driven by gas turbines or electric motors using NG as fuel. At most 90% of the entering feed gas to a modern LNG plant is shipped as exported LNG and 10% of the gas is consumed to produce the required shaft work to liquefy the remaining NG. High operating and capital cost of an LNG plant opens a challenging field for more investigation in refrigeration cycle and optimal configuration of compressors to reduce cost.

Obtaining the best refrigeration system configuration has caused many attentions due to its economic importance. Barnes and King [3] investigated the problems of synthesizing refrigeration cycles and provided a two-step approach to identify optimum cascade refrigeration systems. In the first step, a limited number of promising choices for configurations and design parameters were identified using graph decomposition principles. To minimize the cost of the configuration, the problem was represented as a network. Later, Cheng and Mah [4] proposed an interactive procedure for synthesizing refrigeration systems incorporating all the refrigeration features identified by Barnes and King. The refrigerants participating in a cycle were selected based on their allowable operating temperature range and the temperature of the process streams to be cooled. Townsend and Linnhoff [5] and Linnhoff and Dhole [6] used a set of qualitative guidelines based on pinch technology and exergy analysis for placing heat engines and heat pumps to minimize utility consumption. Aspelund [7] proposed a methodology based on pinch analysis to utilize pressure based exergy for sub-ambient processes, such as LNG. Shin et al. [8]proposed a mixed integer linear programming (MILP) formulation for optimizing boil-off gas (BOG) compressor operations in an LNG re-gasification terminal, and Del Nogal [9] presented an optimization framework for the design of mixed refrigerant cycles which was suitable for LNG.

These methods are general in applicability and share some heuristics to find number of pressure levels, intermediate stages and partition temperature, besides focus has been placed on the process optimization of only a specific part of LNG plant and not on the cascade configuration. When applying these approaches to complex multistage refrigerant cycles the shortcoming of these methods arises. The cascade does not converge as a result of both non-linearity in problem formulation and explosion of integer variables. To overcome this problem, a stepwise procedure has been introduced that the main parameters of a refrigeration cascade like partition temperature and pressure level are firstly determined and in the next step the refrigeration configurations and features are decided. CC and EGCC diagrams are added to analysis to give a better conceptual insight to the designer.

The complex nature of the heat and material balance equations in multi stream heat exchangers (MSHXs) and non-linearity of physical properties of natural gas and refrigerant mixtures makes computation of the model highly non-linear, which leads to use MINLP mathematics. In this paper a new method is introduced to find optimal synthesis of an LNG plant by mounting mixed integer non-linear programming on a superstructure and applying several industrial heuristics. MINLP method is a powerful tool for decision making problems and the new procedures applies it to determine the best compression configuration for the refrigeration cascade.

THEORETICAL PRINCIPLES OF RE-FRIGERATION AND LNG SYSTEMS

In NG liquefaction process, acid gases and mercaptans are removed from sour NG. Cascade refrigeration is required to reach very low temperatures. A simplified cascade refrigeration cycle for mega scale LNG plant consists of three sub-cycles, each using a different pure refrigerant, (Fig. 1).

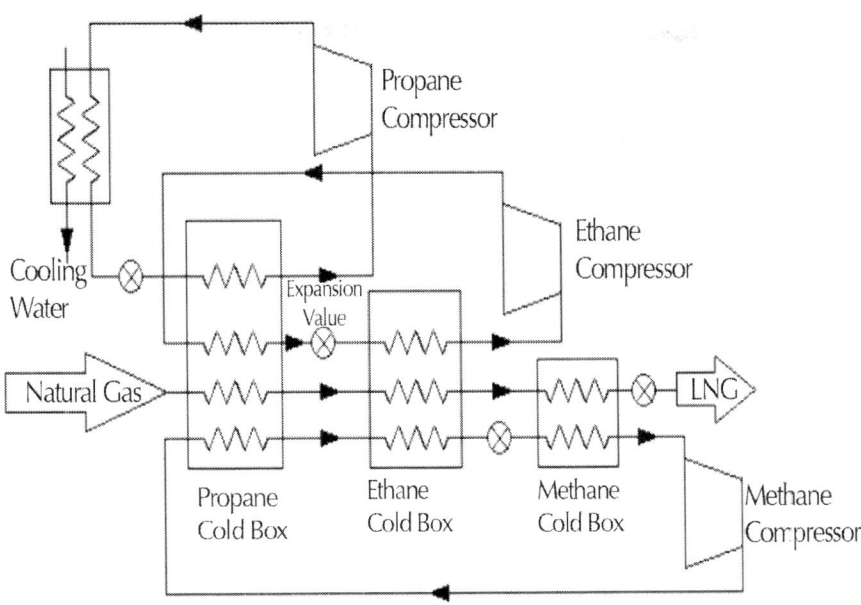

Figure 1: Schematic of cascade refrigeration cycle.

Only one stage for each cycle is shown for simplicity but in real industrial cycles 2 or 3 pressure stages are available by using expansion valves and each stage shall have its own pre-saturator, economizer, de-superheater, *etc.* In the first cycle, propane leaves the compressor at high temperature and pressure and enters the condenser where the cooling water or air is the external heat sink. The condensed propane then enters the expansion valve where its pressure is decreased to the evaporator pressure and the temperature of hot streams decreases to − 40 °C. As the natural gas and methane are cooling down and ethane of lower cycle is condensing, the liquid refrigerant propane evaporates. Propane leaves the evaporator as superheated vapor and enters the compressor, thus completing the loop. The condensed ethane in the middle cycle expands in the expansion valve and evaporates as methane condenses and natural gas is further cooled and liquefied. In ethane cycle, temperature of hot streams decreases to − 100 °C. Finally, methane expands and then evaporates as natural gas is liquefied and subcooled to − 160

°C. As methane enters the compressor to complete the loop, the pressure of LNG is dropped in an expansion valve to the storage pressure [10].

Many refrigeration features are available which can be mounted over simple refrigeration cycles. These options reduce required compression shaft work. A cascade refrigeration system and its P–h diagram are shown in Fig. 2. The lower cycle absorbs heat at temperature levels 1–4 and rejects condensation heat to the upper cycle at temperature levels of 2–3. The upper cycle absorbs rejected heat from the lower cycle by operating at evaporation levels of 5–8, which is colder than levels 2–3. Finally, the heat in the upper cycle is rejected at levels 6–7 to external heat sinks like cooling water and air cooling systems.

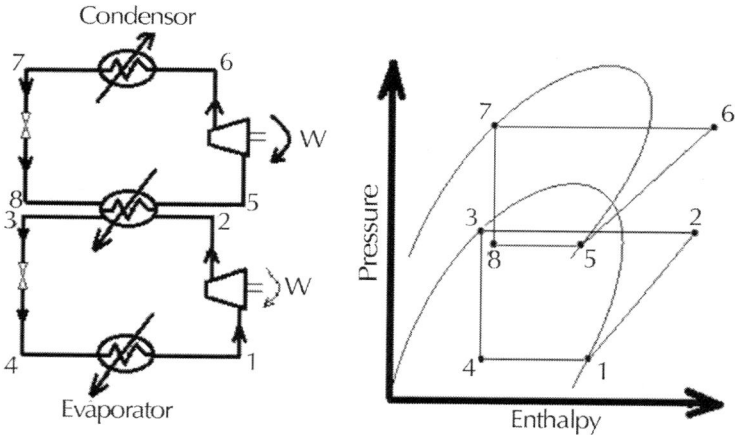

Figure 2: A simple cascade refrigeration system diagram.

The reasons for using this kind of cascade refrigeration systems are two-folds. First, there are no single refrigerant in a single cycle to cover all temperature range of refrigeration. Second, in terms of energy consumption, using a single refrigerant for the whole refrigeration demand may consume more shaft work than using multiple refrigerants. Some basic features of refrigeration in a superstructure are described inSection 2.1.

Refrigeration Features for Shaft Work Saving

For a refrigeration system, it is possible to improve its performance by using following design options [11], as shown in Fig. 3:

- Economizer: as presented in Fig. 3(a). In an economizer, the condensed refrigerant is flashed to an intermediate pressure, where the flash vapor is returned to the suction of the compressor and the remaining liquid is further expanded to a lower temperature. As a result, the amount of vapor f owing through the lower pressure part is reduced, thus saving shaft work.

- Aftercooler: as seen in Fig. 3(a). With this option, the super reated refrigerant vapor is cooled down after compression by other available heat sinks before further compression. This causes reduction of required shaft work and the after-cooling duty. Also, after-coolers provide the opportunity of heat integration between refrigeration systems and processes.

- Presaturator: as indicated in Fig. 3(c). A presaturator has a similar structure as that of an economizer, but the partially compressed refrigerant vapor is presaturated in the flash vessel with the expense of evaporating some part of the refrigerant liquid from the corresponding economizer. This decreases the temperature of the refrigerant vapor entering the next stage of compressor, and saves shaft work. On the other hand, pre-saturation may have two drawbacks: (1) it requires a higher refrigerant flow rate which may cause more compression shaft work and (2) both economizer and presaturator, add an intermediate pressure level, which may cause an increase in capital cost for compressors. Several small compressors can be more expensive than a single large compressor, even though the total shaft work requirement is reduced.

- Desuperheater: as displayed in Fig. 3(c). Using a desuperheater in the final stage, the superheated refrigerant vapor is pre-cooled after compression by a warmer heat sink before entering the condenser. This adds the possibility of heat integration to processes.

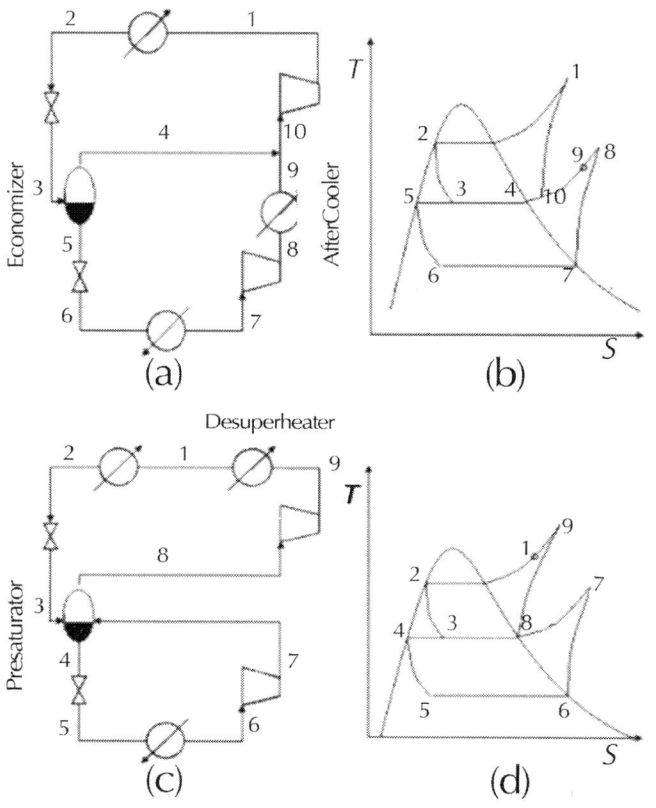

Figure 3: Refrigeration system design options in a cycle.

Determining Compression Configuration Scenarios for Refrigeration

There are many refrigeration configurations and the optimal synthesis of the cascade should be determined for the lowest capital and operating cost. Main decisions for a refrigeration cascade include compression configuration, number of stages in each compression section and refrigeration base temperature that are shown in Fig. 4.

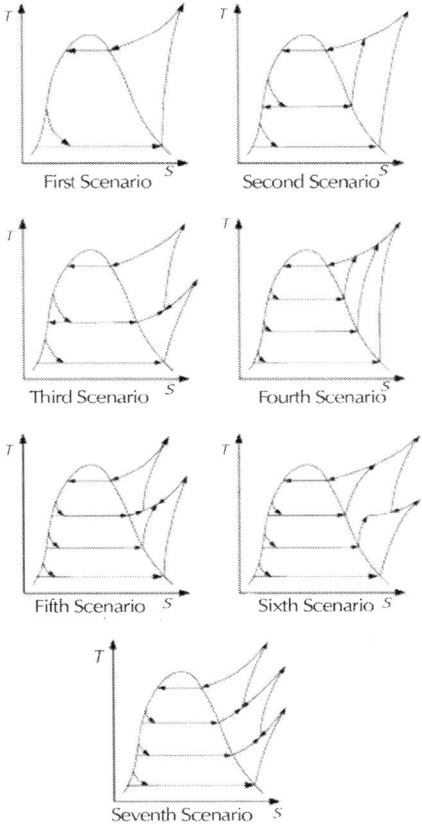

Figure 4: Compression configuration scenarios.

After deciding on these major components, a refrigeration superstructure could be established and then all other refrigeration features could be mounted on the obtained configuration and the superstructure will be optimized.

An important parameter to be determined in a cycle is the number of pressure levels and the associated pressure of it. Three possibilities are considered for pressure levels in a cycle. Any cycle may include one, two, or three pressure levels. Fig. 4 represent different configurations of compressor which is a binary variable in our superstructure modeling. The maximum number of sections (volutes which can mechanically increase pressure from inlet to discharge) in a compressor set is 7 and they will be placed between pre-specified pressure levels as shown in Fig. 4.

Seven scenarios for compression configuration are considered in Fig. 4. The first scenario is a simple refrigeration cycle with no inter-stage. The second scenario has two compression stages which results in 2 pressure levels. Vapor refrigerant from the lower level is compressed to the highest pressure and the vapor from the medium pressure level is compressed to higher pressure and is mixed with the other stream. In the seventh scenario there are three stages with three pressure levels. Vapor refrigerant from the lower level is pressured to the second pressure level and is mixed with the incoming refrigerant vapor. The mixture is pressured up to the third stage and is mixed with refrigerant vapor from the third level. All mixed refrigerant are compressed to the highest pressure and the heat load of the superheated vapor is rejected either to the higher cycle or to the ambient heat sink. All other scenarios could be defined similarly.

The above scenarios are mathematically modeled in Section 2.5 and an MINLP solver can find the best scenario which minimizes cascade shaft work and capital cost. The different refrigeration features like pre-saturator, economizer and desuperheater shall be mounted over the selected scenario and therefore energy consumption is further reduced.

Technical Heuristics to Find the Best Refrigeration Cascade in LNG Industry

Dealing with complicated problems like multistage refrigeration cascade, some industrial practices and constraints can help to achieve a realistic and applicable design. The below items are some practical guidelines which are used in the design of cascade systems:

- The size of LNG plant dictates the complexity of the design. LNG plants with capacities less than 1 million ton per annual (MTPA) only use one cycle and the designer should avoid a cascade design. This single cycle can be a multistage cycle and all refrigeration features like economizer, presaturator, and re-boiler, could be applied. When the LNG plant size increases, it is logical to use two or three cycle in a cascade and same features used in single cycle could be used on it. [12].

- The lowest temperature of natural gas in a cascade is dictated by the required composition of produced LNG. LNG quality is determined by the main market which it shall be exported, for example the European market requires lower HHV (~ 970 MMBTU/SCF) and East Asia requires higher HHV (~ 1100 MMBTU/SCF). When the main market for plant is determined, the specification of product is known and the lowest required temperature will be found. This temperature shall dictate suction pressure of the lowest cycle compressor. [12].

- LNG plants are the largest vapor recompression cycle in the world. With regard to the pressure ratio and flow rate, best choice for compressors is the centrifugal ones. Compressor manufacturers build compressors which have at most seven stages as a normal practice and compression ratio of each stage is around 1.7. Pressure levels in the cycle are determined by multiplying base pressure to this ratio powered by number of sections between corresponding pressure levels [12].

- Each cycle transfers heat load of process stream and work of compressor to the upper cycle. The returning refrigerant from the higher cycles should be fully condensed, as the main heat rejection usually occurs during condensation [13].

- Partition temperatures are a very important characteristic of any LNG cascade. It divides compression load of refrigeration and temperature range where cooling occurs. There are two guidelines to place partition temperature between each cycle:

- Superheated refrigerant that is discharged from lower cycle to the upper cycle should be returned in liquid phase. [13]

- Discharge temperature of vapor stream of each compressor shall not exceed 135 °C [14].

Design Methodology to Find Optimal Pressure Level and Intra-Cycle Partition Temperature Placement by Using Grand Composite and Exergetic Grand Composite Curves

Refrigeration cascade design starts from the lower cycle to the upper cycle, as there is no external heat load from any cycle to the lowest cycle. At first step as shown in Fig. 5(a), the cooling demand curve is drawn in a grand composite curve (GCC), and then an initial partition temperature that divides cooling load between the lower and the upper cycles is assumed. Further, the refrigeration load of lower cycle is met, and the heat is rejected to the upper cycle and the GCC is updated as indicated in Fig. 5(b). The effect of introducing a pressure level and refrigeration option in second cycle is shown in the grand composite curve (GCC) of Fig. 5(c). At last the accumulated heat load is rejected to ambient heat sink.

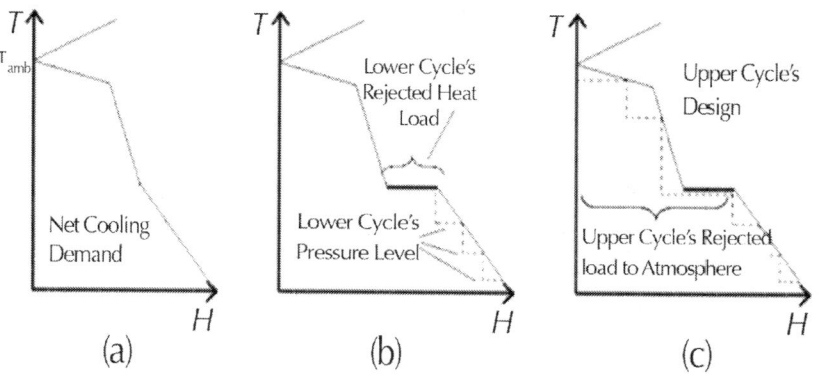

Figure 5: Refrigeration cascade design procedure.

If temperature axis of the GCC diagram is turned to Carnot factor, then exergetic grand composite curve (EGCC) is obtained [6]. Introducing any new pressure level or refrigeration feature in a cycle results in lower exergy loss and compressor shaft work as displayed in Fig. 6.

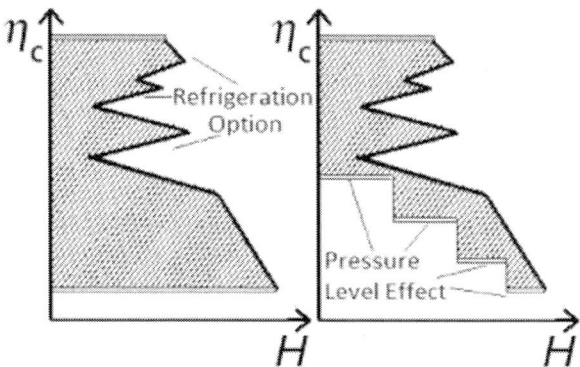

Figure 6: Effect of new refrigeration features and pressure level on shaft work reduction.

Fig. 6 enables us to evaluate the effect of different design options in the refrigeration system quickly and visually. EGCC diagrams of a cascade help the designer to estimate required compression shaft work. EGCC guides design procedure to find the best compression configuration and partition temperature by minimizing the area encircled between utility line and EGCC diagram.

By using the developed theoretical principles and these heuristics, mixed integer non-linear mathematics can model heat-material balance of refrigeration cascade which includes decision making parameters like existence or non-existence of pressure levels and compression configuration, selects between economizer and presaturator and minimizes capital and operating cost of plant. The MINLP method is used where logical selections or different sets of equations should be applied to different design scenarios. The MINLP method as a decision-making tool helps to determine the best configuration, which is discussed inSection 2.5.

Mixed Integer Non-linear Programming Model in Refrigeration Systems

MINLP is a form to model problems in which discrete variables are restricted to values of 0 and 1 and represent certain decisions which are necessary to deal with continuous variables [15].

In a refrigeration superstructure a framework is developed by allowing a bypass model to take effect when a given option is eliminated from the superstructure. Such model has the following form:

$$\text{Min} Z = \sum_i \sum_k C_{ik} + d_y^{\text{T}}$$

$$\text{s.t.} \qquad g(x) \leq 0$$

$$r(x) + Dy \leq 0$$

$$Ay \geq a$$

$$\begin{bmatrix} Y_{ik} \\ h_{ik}(x, C_{ik}) \leq 0 \\ C_{ik} = \gamma_{ki} \end{bmatrix} \vee \begin{bmatrix} -Y_{ik} \\ B^i x = 0 \\ C_{ik} = 0 \end{bmatrix}$$

$$\Omega(Y) = \text{True}$$

$$x \in R^n, c \in R^m, y \in \{0,1\}, Y \in \{\text{True, False}\}^m.$$

$$(1)$$

Y_{ik} are Boolean variables that determine whether a given term (heat-material balance) in a disjunction is true [$h_{ik}(x,c_{ik}) \leq 0$] or false [$h_{ik}(x,c_{ik}) \geq 0$]. x and C_{ik} are continuous variables, the latter being used to model annualized costs associated with each disjunction and $\Omega(Y)$ are logical relations assumed to be in the form of propositional logic involving only the Boolean variables. In $g(x) \leq 0$, 0 represents thermodynamic and industrial constraints that are valid over the entire search space while the disjunction $k \in SD$ states that at least one subset of constraints $h_{ik}(x,c_{ik}) \leq 0$, $i \in D_{ik}$ must be hold (*i.e.*, presaturator and an economizer cannot co-exist simultaneously in the same stage). Y_{ik} are auxiliary variables that control the part of the feasible space in which the continuous variable x lie, while the logical condition $\Omega(Y)$, expresses relations between the disjunctive sets.

$r(x) + Dy \leq 0$ represents the general mixed-integer algebraic formulations in which the original disjunctions are transformed into algebraic equations. $Ay \geq a$ is a set of integer inequalities and dy^T are linear cost terms. This form is more flexible than r gorous modeling. Disjunctive programming can be used as a basis to formulate a mixed-integer program with 0–1 variables [15].

Fig. 7 and Fig. 8 describe a superstructure model for a cycle between level k and the above level.

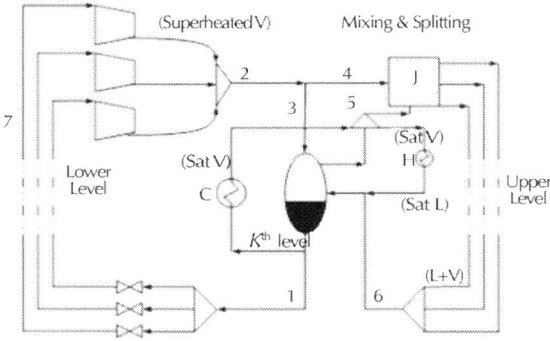

Figure 7: kth level of refrigeration cycle superstructure. V-vapour; L-liquid;C-condenser; H-heater; J-junction.

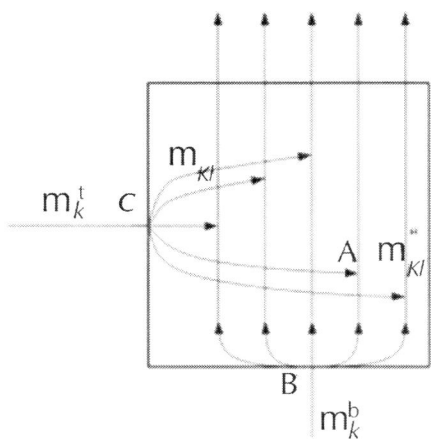

Figure 8: Mixing and dividing streams in junction J of Fig. 7 superstructure.

Consider a cycle operating between levels k and l, where k is below l. Key variables of interest are the refrigerant flow rates $\dot{m}kl$ within cycles operating between levels k and l. H_{kl} is the rejected heat to the cycle operating between levels k and l by level k, the work input W_{kl} to the cycle, and the enthalpies hk^{cp}, and hkl^{out}.

The modeling equations are heat and material balances at various mixing and dividing points in the configuration.

$$\sum_{l:(k,l)} \dot{m}_{kl} + \dot{m}_k^i = \sum_{l:(l,k)} \dot{m}_{kl} + \dot{m}_k^t \tag{2}$$

$$\sum_{l:(k,l)} \dot{m}_{kl} * h_{kl}^{in} + \dot{m}_k^i * h_k^{cp} + \sum_{l:(l,k)} H_{kl} = \sum_{l:(l,k)} \dot{m}_{lk} * h_k^{liq} + \dot{m}_k^t * h_k^v$$
$$+ \sum_{l:(k,l)} H_{lk}. \tag{3}$$

Heat material balance at Points B and C is:

$$\dot{m}_k^b = \sum_{l:(k,l)} \dot{m}_{kl}'' \tag{4}$$

$$\dot{m}_k^t = \sum_{l:(k,l)} m_{kl}'. \tag{5}$$

Heat material balance at junction A is:

$$\dot{m}_{kl}' + \dot{m}_{kl}'' = \dot{m}_{kl} \tag{6}$$

$$\dot{m}_{kl}' * h_k^{vap} + \dot{m}_{kl}'' * h_k^{cp} = \dot{m}_{kl} * h_{kl}^{out}. \tag{7}$$

The linking relation between the absorbed heat by a cycle and the refrigerant flow rate is:

$$H_{kl} = \dot{m}_{kl} * \left(h_{kl}^{out} - h_{kl}^{in} \right). \tag{8}$$

The relation between the flow rates of all cycles operating between levels below k and level k and the absorbed energy by them to level k is as follows:

$$\sum_{l:(l,k)} (H_{kl} + W_{kl}) = \sum_{l:(l,k)} \dot{m}_{kl} \left(h_k^{cp} - h_k^{liq} \right).$$

(9)

Finally, the defining relation for the compression work is:

$$W_{kl} = \sum_{l:(l,k)} \dot{m}_{kl} \left(h_k^{out} - h_k^{in} \right).$$

(10)

The fact that, when an option is omitted its related constraints or equations are sufficiently relaxed or unbounded which makes the minimization become robust and computationally efficient. Much less time is wasted to calculate and converge those virtually non-existing equations. In each intermediate pressure level of a cycle existence or non-existence of refrigeration features that was discussed in Section 2.1 and other constraints is modeled by the following mathematical equations.

Vapor–liquid Heat Exchangers Modeling

In a V–L heat exchanger, incoming saturated refrigerant liquid exchanges heat with refrigerant vapour after evaporation. Fig. 9 shows the situation when a V–L heat exchanger is placed in the kth level of a refrigeration cycle. Discrete modeling is applied to describe the existence of a V–L heat exchanger in the kth level:

$$\begin{bmatrix} y_k^V \\ y_k^L + y_{k-1}^L + \ldots + y_1^L = 1 \\ T_k^{in} = T_k^e + \dfrac{q_k^v}{\dot{m} \cdot C_p} \\ T_k^{in} \leq \sum_{i=k-1}^{1} y_i^L \cdot T_i^e - \Delta T \end{bmatrix} \vee \begin{bmatrix} y_k^V \\ T_k^{in} = T_k^e \end{bmatrix}$$

(11)

yk^V and $-yk^V$ denote the existence and non-existence of a vapour heat exchanger at the compressor's suction vapour line, respectively. When a vapour heat exchanger is placed, there must be a liquid heat exchanger so that the energy balance is satisfied.

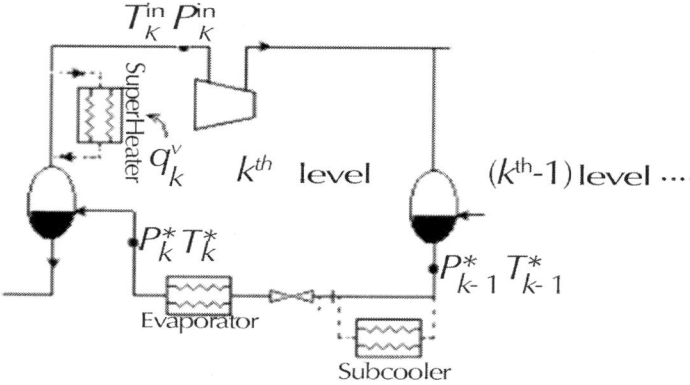

Figure 9: A V-L heat exchanger in the *k*th level.

Any liquid heat exchanger with saturated temperature higher than or equal to $Tk_{-1}{}^e$ is possible, but only one is needed. This condition is described by the first constraint of the left hand side. The third constraint guarantees validity of heat transfer processes. On the other hand, if a vapour heat exchanger is not placed, the condition of inlet stream of the compressor is simply set to equal to outlet conditions of the evaporator. This discrete model is transformed into a mathematical model by using "Big-M" transformation [15]:

$$\left(y_k^L + y_{k-1}^L + \dots + y_1^L\right) + M\left(1 - y_k^V\right) \geq 1$$

$$\left(y_k^L + y_{k-1}^L + \dots + y_1^L\right) + M\left(1 - y_k^V\right) \leq 1$$

$$T_k^{in} = T_k^e + \frac{q_k^v}{\dot{m} \cdot C_p} \cdot y_k^V$$

$$T_k^{in} \leq \sum_{i=k-1}^{1} y_i^L \cdot T_i^e - \Delta T + M\left(1 - y_k^V\right).$$

$$(12)$$

Discrete Modeling of after Cooler

An aftercooler is considered when there is a suitable external heat sink, such as air, cooling water or cold utilities. The placement of an aftercooler in the *k*th level of a refrigeration cycle is shown in Fig. 10.

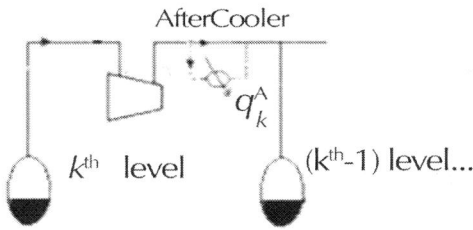

Figure 10: Placement of an aftercooler in the kth level.

Discrete modeling formulation for placement of an aftercooler is:

$$
\begin{bmatrix}
y_k^A \\
T_k^{out,A} = T_k^{out} - \dfrac{q_k^A}{\dot{m} \cdot C_p} \\
T_k^{out,A} = T_k^{Amb} + \Delta T
\end{bmatrix} \vee \begin{bmatrix} y_k^A \end{bmatrix}
\tag{13}
$$

yk^A and $-yk^A$ denote the existence and non-existence of an aftercooler in the kth level. Decrease of temperature after the aftercooler is calculated by the second constraint. The third constraint guarantees the validity of the heat transfer process. If there is no aftercooler being placed, nothing needs to be done. Big-M transformation of this model is:

$$
\begin{bmatrix}
T_k^{out,A} = T_k^{out} - \dfrac{q_k^A}{\dot{m} \cdot C_p} \cdot y_k^A \\
T_k^{out,A} + M\left(1 - y_k^A\right) \geq T_k^{Amb} + \Delta T
\end{bmatrix}
\tag{14}
$$

Discrete Modeling of Pre-saturator and Economizer

When an economizer is placed, the condensed refrigerant is flashed to an intermediate pressure, where the flash vapour is returned to the suction of the compressor. When a pre-saturator is installed, the compressed refrigerant vapour is presaturated in the flash vessel with the expense of evaporating part of the refrigerant liquid from the previous level. The dotted line in Fig. 11 represents a pre-saturator option.

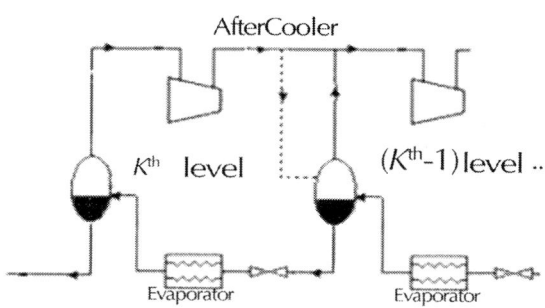

Figure 11: Presaturator option (dotted line).

The reason to discuss two options together is that they cannot co-exist simultaneously in the same stage. For installing a pre-saturator, the model is:

$$\begin{bmatrix} \dot{m}_{k-1}^{\mathrm{cin,P}} = \dot{m}_{k-1}^{\mathrm{cin}} + \dfrac{\dot{m}_k^{\mathrm{cout}} \cdot C_p \cdot \left(T_k^{\mathrm{cout}} - T_{k-1}^{\mathrm{e}}\right)}{h_{k-1}^{\mathrm{vap}}} \end{bmatrix} \vee \begin{bmatrix} -y_k^{\mathrm{P}} \end{bmatrix}$$

(15)

And for installing an economizer, the model is:

$$\begin{bmatrix} y_k^{\mathrm{C}} \\ q_k^{\mathrm{E}} = 0 \end{bmatrix} \vee \begin{bmatrix} -y_k^{\mathrm{C}} \end{bmatrix}$$

(16)

Since a pre-saturator and an economizer are not possible to co-exist in the same level, the following constraint is required:

$$y_k^{\mathrm{P}} + y_k^{\mathrm{C}} \leq 1$$

(17)

The complete mathematical formulations that describe selection of two options are expressed as:

$$\dot{m}_{k-1}^{cin.P} = \dot{m}_{k-1}^{cin} + \frac{\dot{m}_k^{cout} \cdot C_P \cdot \left(T_k^{cout} - T_{k-1}^e\right)}{h_{k-1}^{vap}} \cdot y_k^P$$

$$q_k^E \leq M\left(1 - y_k^C\right)$$

$$q_k^E \geq -M\left(1 - y_k^C\right)$$

$$y_k^P + y_k^C \leq 1. \tag{18}$$

The introduced procedure is incorporated in a program (LNG-Pro) which integrates VBA, Refprop and Excel MINLP Solver to automate the methodology. LNG-Pro starts optimization from the lowest cycle and after meeting the heat-material balance the MINLP solver determines which integration of above features m nimizes compression shaft work. The developed program starts optimization from the lowest cycle and after meeting the heat-material balance (HMB) and above constraints. Both rejected process heat load and load of compressor are shifted up to the highest and middle cycles. Intermediate pressure and flow rates are constrained not to allow temperature cross in MSHXs. The procedure is accomplished in two steps. In first step, a simple model is established by omitting refrigeration features and scenarios and is only limited to have a sub-cooler as a refrigeration feature. The goal of this stage is to determine the main parameters of a cryogenic cascade like partition temperature and pressure level. This procedure is shown in Fig. 12. The objective function is to minimize compression shaft work. The program tries to find partition temperature and a set of pressure levels (condensing and evaporating pressures) that can give the best match between hot and cold composite curves by adjusting refrigerant flow rate. If the search is successful, then pressure levels and/or the refrigerant flow rate are reduced progressively. The procedure is terminated when no set of valid refrigerant flow rate and pressure level can be found and temperature crosses occur in MSHXs.

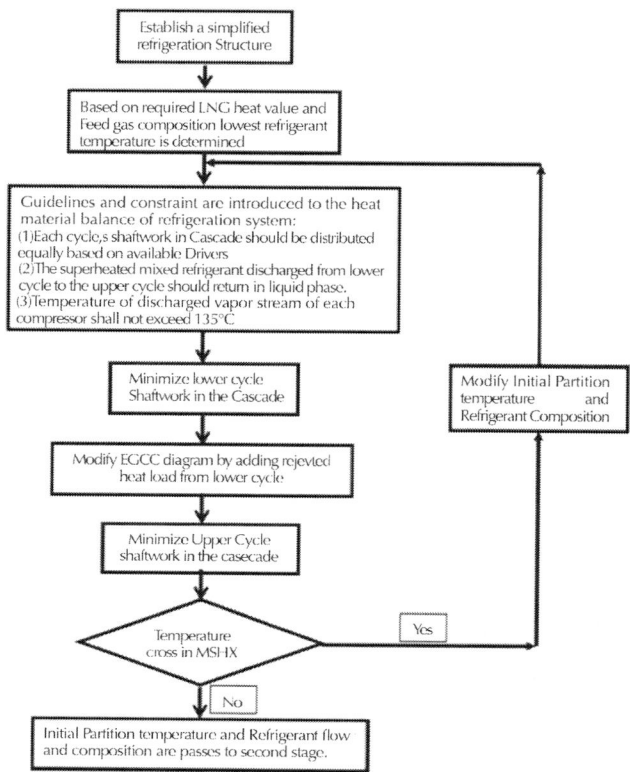

Figure 12: First step of design procedure to find initial partition temperature and refrigerant flow.

The procedure in the second step determines refrigeration scenarios and features like sub-cooler, after cooler, Presaturator and economizer is shown in Fig. 13.

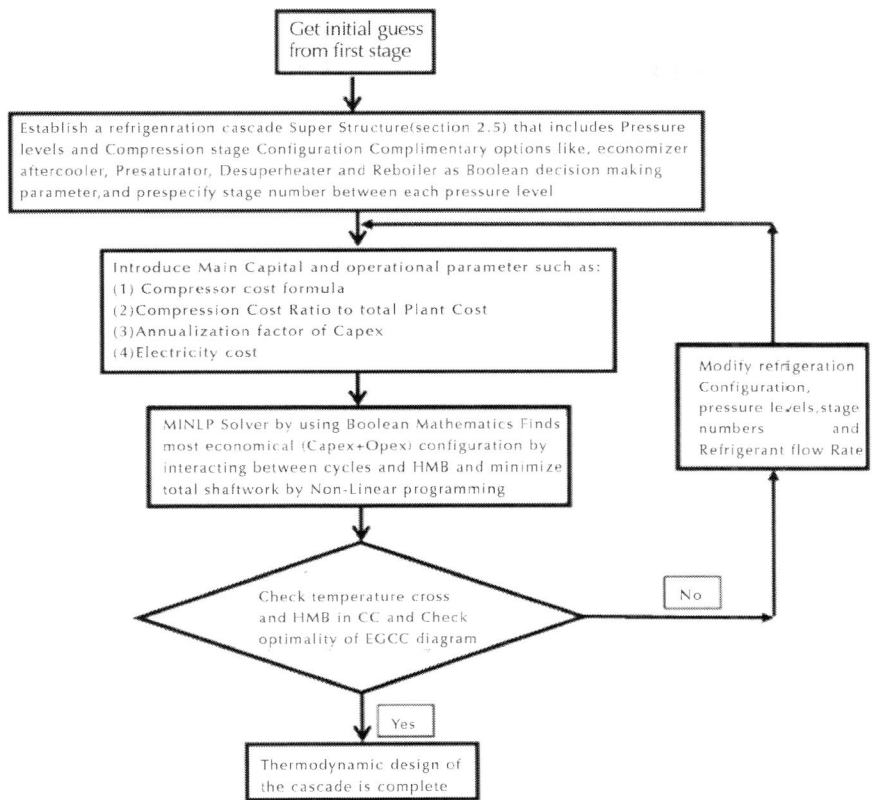

Figure 13: Second step of design procedure to find best compression configuration and refrigeration features by minimizing CAPEX and OPEX.

The objective function for second step is to minimize OPEX and CAPEX of the plant. The program tries to find a set of refrigeration scenarios by adjusting refrigerant flow rate and/or refrigeration scenarios, the procedure terminates when either the refrigerant flow rate is too small or temperature crosses occur in MSHXs.

The procedure is applied on a mega scale LNG plant with the liquefaction capacity of 25 million·m³ per day of natural gas that is equal to the upstream production rate of one standard phase of Iran's south pars gas field. The design procedure shows the method's ease of use, flexibility and applicability.

SAMPLE LNG PLANT DESCRIPTION

Suppose that there is a gas field with a capacity of around 90 BCM and lifecycle of LNG plant is 25 years. This amount of gas equals to an LNG plant with the capacity of 5.4 MTPA of LNG. Treated natural gas composition is propane 3%, ethane 5%, methane 90%, and nitrogen 2% on mole basis and its pressure is 9 MPa. A set of multi stream heat exchanger (MSHX) with an approach temperature of 5 °C, compressors with isentropic efficiency of 82% and ambient temperature of 300 K is available. As the size of the plant is big enough, it could be partitioned into three different cycles [12]. Evaporation rate after each expansion valve is assumed to be 5%, 10% and 15% for the lowest, the middle and the highest cycle for the initial guess, and the base pressure is 0.225 and 0.115 and 0.115 MPa, respectively.

The design starts by establishing a simple refrigeration structure by using methods described in Section 2 to find compression configuration, number of stages in each compression section and refrigeration base temperature.

The nature of this phase of design is decision making and existence or non-existence of a compression configuration or pressure level introduces 2 different sets of integer variables. Natural gas and heat-material balance equations in the superstructure model are all non-linear. There are thousands of refrigeration configurations that can meet an LNG cycle thermodynamically, but these configurations must be constrained by many industrial limits and operational and economical parameters to tailor the best multistage cascade cycle for an LNG plant should be considered.

In the aforementioned superstructure all expenses have been annualized. The electricity cost for compressor driver is assumed to be 0.06 $·(kW·h)$^{-1}$ and the cost of compressor as the main single component of a liquefier is estimated to be: 740 [shaft work (KW)] + 612630, [16]. The cost of compressor for such a plant is estimated to be around 12% of the total plant cost [16].

The LNG-Pro Program starts optimization by applying the aforementioned MINLP formulas in Section 2.5and algorithm described in Fig. 12.

Result summary of this step is represented in Table 1.

Table 1: Optimal compression configuration and refrigerant evaporation pressure level and minimized OPEX and CAPEX cost of plant

Cycle position	Partition temp/K	Selected scenario	Pressure levels/MPa	Compressor section configuration	Shaft work / MW	Annualized capital cost × 10⁶/USD	Annual operating cost × 10⁶ /USD
Higher	239	5th	0.115–0.195–0.332–1.633	1–1–3	127.38	1316.6	63.28
Middle	195	7th	0.115–0.195–0.332–1.633	1–1–3	62.86	651.1	31.23
Lower	122	7th	0.225–0.65–1.879–5.43	2–2–2	37.2	389.5	18.48

Composite curve (CC) and grand composite curve (GCC) of the selected LNG cascade are shown inFig. 14 and Fig. 15. These curves summarize heat load of hot and cold streams in the refrigeration system. Straight lines indicate evaporation temperature of refrigerants at different pressure levels. The area between hot and cold curves is an indicator of irreversibility and exergy loss in the system. These curves help in assuring that the heat material balance is met and no heat cross between hot and cold stream has happened. Fig. 14 shows the pinch point that is the most critical place in our design. For a more flexible and reliable design, the sharp edges in composite curve should be avoided and for saving energy consumption the area between these curves should be decreased.

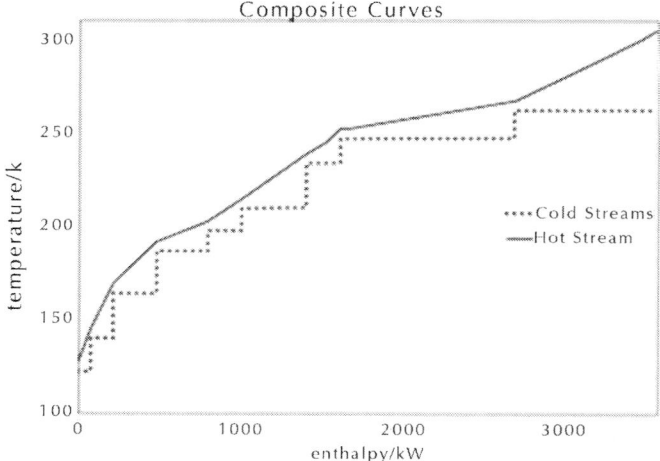

Figure 14: LNG cycle's composite curve after determining main refrigeration configuration.

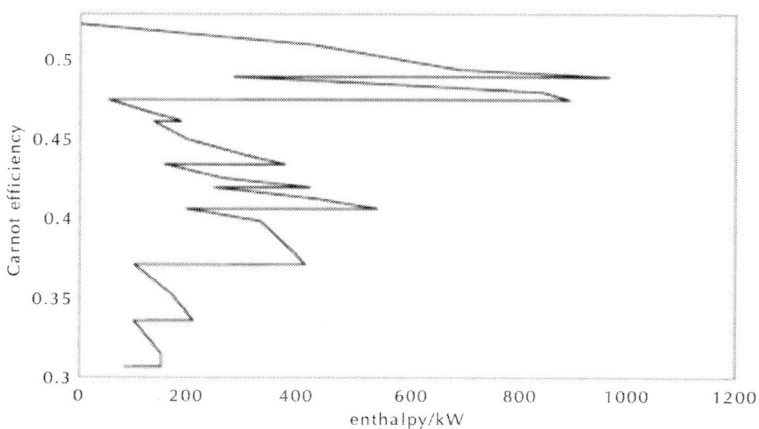

Figure 15: LNG cycle's exergetic grand composite curve after determining main refrigeration configuration.

Fig. 16 shows driving force diagram of optimum refrigeration cycle between cold and hot streams. Straight lines mean constant evaporation temperature of refrigerant.

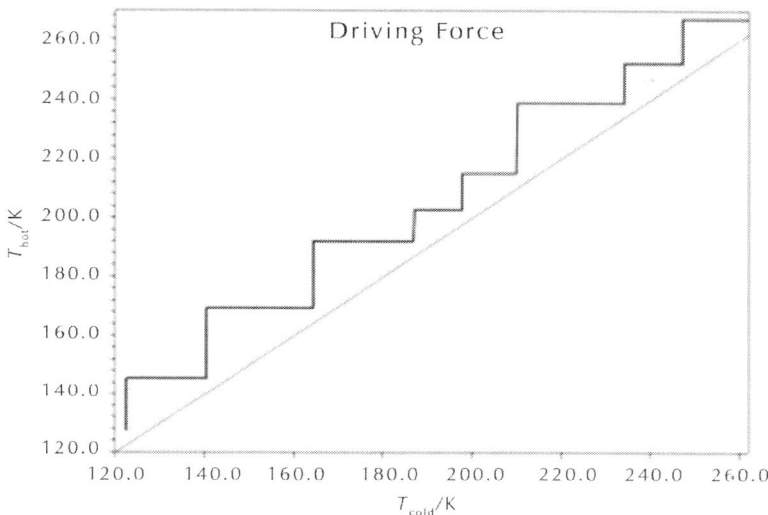

Figure 16: Driving force curve of LNG cycle after determining main re-frigeration configuration.

If the above assumptions are changed, for instance, give more importance to operating cost (higher electricity price) or capital cost, different configuration will be obtained which are compatible to those conditions. Construction cost for such a plant is anticipated to be 2357.24 million $.

Table 1 summarizes base configuration that includes the main parameters of a refrigeration system like pressure levels and compression configuration.

In second step complimentary options such as economizer, aftercooler, pre-saturator and desuperheater are added to this structure. These options in superstructure model are disjuncted by Boolean variables and the problem is modeled by equations found in 2.5.1, 2.5.2 and 2.5.3. This model is solved by MINLP Solver Engine in LNG-Pro Program. LNG-Pro uses the algorithm of Fig. 13 to find the best refrigeration configuration with pure refrigerant.

The final composite curve and grand composite curve and driving force diagrams of the cascade are shown in Fig. 17, Fig. 18

and Fig. 19. The calculated shaft work for the optimized multistage cascade is 206.87 MW which is well below the first step of 227.44 MW. This shaft work reduction is a direct result of mounting more refrigeration features that means more capital cost and a more complicated cycle.

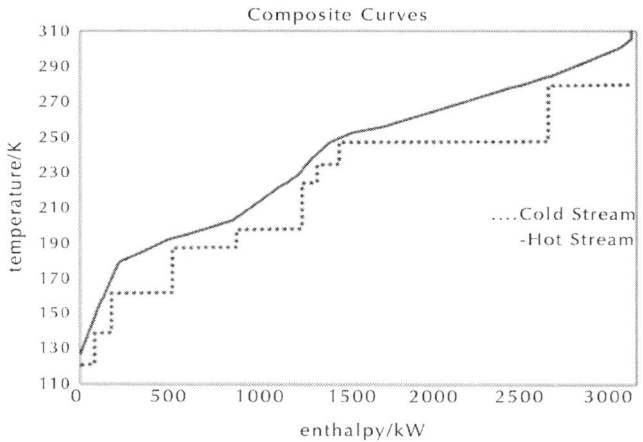

Figure 17: LNG cycle's composite curve including all refrigeration features.

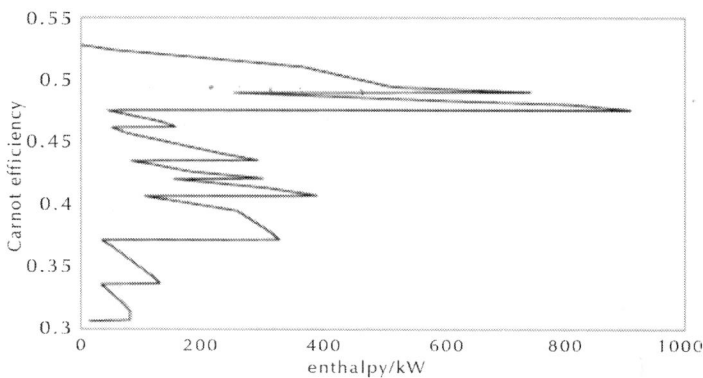

Figure 18: Exergetic grand composite curve of LNG cycle including all refrigeration features.

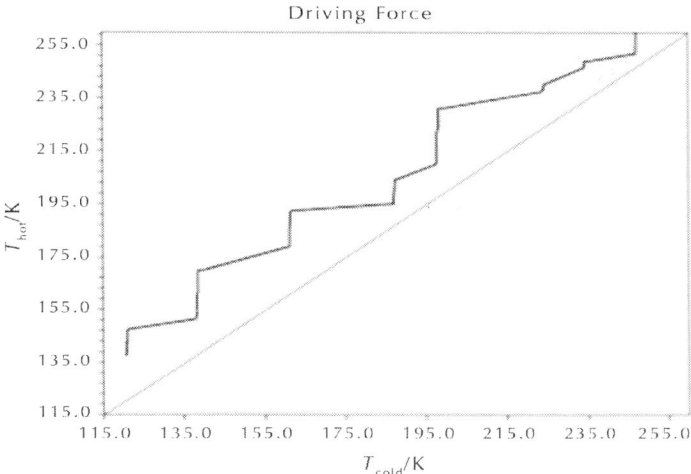

Figure 19: Driving force curve of LNG cycle including all refrigeration features.

The step by step progress in the design procedure is shown by comparing composite curve (CC) and grand composite curve (GCC) of Fig. 14 and Fig. 15 with Fig. 17 and Fig. 18. The introduction of complimentary refrigeration options has increased heat integration of cascade and has resulted in an inclined line ofFig. 19. Driving force of the preliminary cascade (Fig. 17) is reduced in driving force of the final model (Fig. 19). This reduction results in a lower shaft work consumption of the final cascade.

Table 2 summarizes heat and power balance for the optimal cascade cycles.

Table 2: Heat and shaft work summary of the final refrigeration cascade

	NG cooling rate/ kW	Compressor shaft work/ MW	Heat rejected from lower to upper cycle/kW	Refrigerant flow rate (ratio to NG)/ $kg \cdot s^{-1}$
Lowest cycle	263.96	42.54	500.88	

Middle cycle	293.93	60.19	1092.81	
Highest cycle	209.29	104.15	1897.88	
Methane				1.5336
Ethane				2.08
Propane				5

Optimum cascade is modeled in Aspen-Plus and results are verified by it. Obtained results from LNG-Pro and Aspen-Plus are compared in Table 3. It shows required coincidence for conceptual modeling of the introduced procedure.

Table 3: Comparison between required shaft work of LNG-Pro and Aspen-Plus

	Final model	Aspen-Plus
First section of lower cycle	12.46	12.54
Second section of lower cycle	46.82	53.41
Third section of lower cycle	175.56	177.79
First section of middle cycle	23.86	25.20
Second section of middle cycle	42.21	48.92
Third section of middle cycle	266.11	268.44
First section of higher cycle	12.10	12.14
Second section of higher cycle	342.73	374.1
Third section of higher cycle	220.06	185.41
Total compression work	1141.91	1157.95

There is around 1% difference between LNG-Pro and Aspen-Plus in total shaft work consumption. This deviation arises from some simplifying assumption in LNG-Pro. The main goal of LNG-pro is to determine compression configuration and refrigeration features by minimizing annualized cost of an LNG plant by conceptual procedures. LNG-Pro does not intend to model refrigeration cycles rigorously which is a task of Aspen-Plus. In LNG-Pro, approach temperature of MSHX is limited to 5 °C which is only applied at two ends of the exchanger, but in Aspen-Plus temperature approach is checked through full length of the exchanger. Simplifying assumptions in LNG-Pro is required to speedup design during decision making stages which causes around 1% deviation from Aspen-Plus.

CONCLUSIONS

A stepwise design procedure for design of complex refrigeration system is introduced in this paper. The introduced procedure solves the time consuming and exhausting mathematical procedures by applying a 2 phase approach to solve major decision making parameters. In first step important parameters like compression configuration, partition temperature and base pressure are determined. In this stage the heuristics are extensively used, which are elaborated to reduce calculation time. The MINLP solver tries to satisfy the major refrigeration requirements, like heat material balance equations, equal compressor load sharing for each cycle, rejecting all latent heat of superheated refrigerant of lower cycle to the upper cycle, and finding the best pressure level in each cycle. In the second step all refrigeration features like sub-cooler, economizer, and presaturator, are mounted on top of the optimum superstructure of the first step and then by minimizing the OPEX and CAPEX of the plant, the optimal synthesis of the LNG plant is found. The first step is a conceptual step to determine the major decisions in a refrigeration cycle and uses simplified assumptions to speed-up calculation, but in the second step a more detailed simulation is used that includes complimentary refrigeration features.

This stepwise procedure is automated in a program (LNG-Pro) and is applied on a 5/4 MTPA LNG plant that uses 209.9 MW for compression shaft work. Energy consumption to liquefy 1 kg of natural gas in the 1st and the 2nd step is found to be 1255.47 and 1141.9 kJ·kg^{-1} respectively. The final design requires 3.3% less energy than normal LNG plants [2] and is modeled and verified in Aspen-Plus.

REFERENCES

1. British Petroleum Statistical Review of World Energy2004 to 2013. 20–22.

2. A. Mortazavi, C. Somers, Y. Hwang, R. Radermacher, P. Rodgers, Performance enhancement of propane pre-cooled mixed refrigerant LNG plant, Appl. Energy 93 (2012) P138–P147.

3. F.J. Barnes, C.J. King, Synthesis of cascade refrigeration and liquefaction systems, Ind. Eng. Chem. Process. Des. 13 (1974) 421–433.

4. W.B. Cheng, R.S.H. Mah, Interactive synthesis of cascade refrigeration systems, Ind. Eng. Chem. Process. Des. 19 (1980) 410–420.

5. B. Linnhoff, D.W. Townsend, D. Boland, G.F. Hewitt, B.E.A. Thomas, A User Guide on Process Integration for the Efficient Use of Energy (2007) 164–167 2nd edition.

6. V.R. Dhole, B. Linnhoff, Shaftwork targeting for subambient plants, AIChE Spring Meeting, Houston, April, 1989, 1989.

7. Audun Aspelund, David Olsson Berstad, Truls Gundersen, An extended pinch analysis and design procedure utilizing pressure based exergy for subambient cooling, Appl. Therm. Eng. 27 (16) (2007) 2633–2649.

8. M.W. Shin, D. Shin, S.H. Choi, E.S. Yoon, C. Han, Optimization of the operation of boiloff gas compressors at a liquefied natural gas gasification plant, Ind. Eng. Chem. Res. 46 (2007) 6540–6545.

9. F. Del Nogal, J.K. Kim, S. Perry, R. Smith, Optimal design of mixed refrigerant cycles, Ind. Eng. Chem. Res. 47 (22) (2008) 8724–8740.

10. Mehmet Kano&glu, Exergy analysis of multistage cascade refrigeration cycle used for natural gas liquefaction, Int. J. Energy Res. 26 (2002) 763–774.

11. G. Lee, Optimal Design and Analysis of Refrigeration Systems for Low Temperature ProcessesPhD Thesis UMIST, Department of Process Integration, Manchester, UK, 2001.

12. Yoga P. Suprapto, LNG & The World Of Energy (1th Edition), Jakarta, vol. IV, Chapter 30, (2007). 55.

13. G. Venkatarathnam, Performance of an auto refrigerant cascade refrigerator operating in gas refrigerant supply (GRS) mode with nitrogen–hydrocarbon and argon– hydrocarbon refrigerants, J. Cryogenics 49 (2009) 350–359.

14. Wilbert F. Stoecker, Industrial Refrigeration Handbook, 4th ed. McGraw-Hill, 2004. 119.

15. Jan Oldenburg, Wolfgang Marquardt, Disjunctive modeling for optimal control of hybrid systems, Comput. Chem. Eng. 32 (10) (2008) 2346–2364.

16. Max Peters, Klaus Timmerhaus, Plant Design and Economics for Chemical Engineers, 5th edition, 2005. 319

Thermodynamic Analysis and Optimization of a Solar-Powered Transcritical CO_2 (Carbon Dioxide) Power Cycle for Reverse Osmosis Desalination Based on the Recovery of Cryogenic Energy of LNG (Liquefied Natural Gas)

Guanghui Xia, Qingxuan Sun, Xu Cao, Jiangfeng Wang, Yizhao Yu, and Laisheng Wang

Institute of Turbomachinery, School of Energy and Power Engineering, Xi'an Jiaotong University, No. 28, Xianning West Road, Xi'an, Shaanxi 710049, China

ABSTRACT

A solar-powered transcritical CO_2 (carbon dioxide) power cycle for reverse osmosis desalination based on the recovery of cryogenic energy of LNG (liquefied natural gas) is proposed. The system consists of a solar collector subsystem, a transcritical CO_2 power cycle subsystem, a LNG subsystem and a RO (reverse osmosis) desalination subsystem. A thermal storage unit is introduced into the system to guarantee continuous and stable operation of the system. A mathematical model is developed to simulate the system based on several assumptions. The effects of several key thermodynamic parameters on the system performance are examined based on the performance criteria, including daily exergy efficiency, daily mechanical work output and daily fresh work output. Parametric optimization is conducted by genetic algorithm to maximize the daily fresh water output. The results show that the CO_2 turbine inlet pressure has an optimal value to reach the daily maximum exergy efficiency under the given conditions. The daily exergy efficiency could decrease with an increase in condenser temperature, and increase with an increase in mass flow rate of oil and NG turbine inlet pressure. Through parametric optimization, the system can reach the daily exergy efficiency of 4.90% and provide 2537.33 m^3 fresh water per day under the given conditions.

INTRODUCTION

With the society development and the increasing demand of fossil fuels nowadays, environment problem and energy crisis have become more and more prominent. In addition, the shortage of fresh water has also threatened many countries. In order to alleviate the environment impact, the utilization of solar energy, as one of the most promising renewable energy resources has attracted a lot of attentions of researchers and shows a rapid development. On the other hand, great efforts have also been done to develop desalination technologies to solve the lack of fresh water.

Reverse osmosis, as a kind of effective desalination method, has been taken more and more attention and many researchers utilize solar energy to drive RO (reverse osmosis) desalination to produce fresh water. Delgado et al. [1] and Ghermandi et al. [2] had carried out some fundamental explorations in the RO desalination driven by solar energy. For the further research, Delgado-Torres et al. [3] and [4] carried out some detailed analysis for a coupled system of the organic Rankine cycle (ORC), solar parabolic trough collectors and a sea water RO desalination unit. Kosmadakis et al. [5] proposed a reverse osmosis desalination system based on a two stages organic Rankine cycle and examined its feasibility. They[6] and [7] also conducted a simulation and economic analysis of the two stages ORC for RO desalination driven by solar energy to estimate the performance improvement. Kosmadakis et al. [8] and [9] also established low or normal temperature RO desalination system based on ORC driven by solar energy, aiming to identify the performance of low temperature ORC–RO system and search for the thermodynamic properties of organic working fluids. Nafey et al. [10] proposed a combined solar ORC–RO desalination system in order to explore the effect of different energy recovery components from thermal-economic view. Beyond the theoretical investigation, Manolakos et al. [11] conducted the experiment of RO desalination system driven by solar energy under real solar conditions. They [9] and [12] also done some experiments of the ORC and the small RO unit under laboratory conditions.

As mentioned above, the ORC is used to driven RO desalination system to produce fresh water, and organic working fluids are employed to achieve the transform from heat to work. But most of organic working fluids are synthesized, expensive, unstable, toxic, what's more, they have much negative effects on the environment. So, it is urgent to find a working fluid which is environment friendly and has good properties to alleviate the environment impact instead of organic working fluids. Carbon dioxide is just a good choice. CO_2 has good properties of non-flammable, non-toxic, safety than other working fluids and is naturally abundant [13]. In addition, CO_2 has good thermodynamic properties [14]. It can easily reach its supercritical state due to its low critical temperature (critical temperature is 31.1 °C and critical pressure is 7.38 MPa). Supercritical CO_2 in power cycle offers a better temperature profile match between the heat source and working fluid during the absorption heat process, resulting in a decrease in irreversibility loss during the heat transfer process. So the transcritical CO_2 Rankine cycles and the cycles combined with solar energy have been paid more attention [15], [16], [17] and [18]. H. Yamaguchi et al. [19] had proposed a solar powered Rankine cycle using supercritical CO_2 which could not only produce electricity but also provide heat collections of different temperatures. Niu et al. [20] investigated an optimal arrangement of the solar collectors of a supercritical CO_2 based solar Rankine cycle based on experimental investigation. Several other experiments had also been done by Zhang et al. [21] and [22] to examine the feasibility of transcritical CO_2 Rankine cycle driven by solar energy in actual environment and conducted some parameters analysis of transcritical CO_2 Rankine cycle.

But there still exists a problem in transcritical CO_2 Rankine cycle. Due to its low critical temperature of 31.1 °C, it is difficult to enable CO_2 to be condensed by environment. A heat sink with very low temperature is required to condense the CO_2 to liquid state in power cycle. LNG (liquefied natural gas) with a temperature of nearly 112 K (−161 °C) is such a good solution to regard as a heat sink to condense the CO_2. In addition, LNG contains a lot of cryogenic energy during the liquefied process. The use of cryogenic energy of LNG can also improve the thermodynamic performance of the system. Zhang et al. [23] proposed a LNG fueled quasi-combined system of Rankine cycle and Brayton cycle with CO_2 as the working fluid and LNG as the heat sink. Lin et al. [24] had designed a novel transcritical CO_2 Rankine cycle which used exhaust from a gas turbine as its heat source and LNG (liquefied natural gas) as its cold sink. In addition, Song et al. [25] design a solar-driven transcritical CO_2 cycle with LNG as heat sink and the parameter analysis was conducted to examine the effects of key thermodynamic parameters on the system performance. Sun et al. [26] also used the solar-driven transcritical CO_2 cycle with LNG as heat sink to achieve hydrogen production.

In this paper, a solar-powered transcritical CO_2 Rankine cycle for RO desalination based on the recovery of cryogenic energy of LNG is proposed to desalinate sea water by using solar energy. A thermal storage unit is added to alleviate the influence of varied solar radiation. Mathematical model of the system is established and the sensitive analysis of some key parameters is conducted to examine the effects of key parameters on the system performance. In addition, parameters optimization is conducted to obtain the optimal performance of the system.

SYSTEM DESCRIPTION

Fig. 1 illustrates a schematic diagram of the solar-powered transcritical CO_2 power cycle for reverse osmosis desalination based on the recovery of LNG's cryogenic energy. This system employs the solar-powered transcritical CO_2 power cycle based on the recovery of LNG's cryogenic energy to drive reverse osmosis desalination unit, resulting in a production of fresh water from sea water. This overall system consists of a solar collection subsystem, a thermal storage subsystem, a transcritical CO_2 power cycle, a LNG subsystem and RO desalination subsystem.

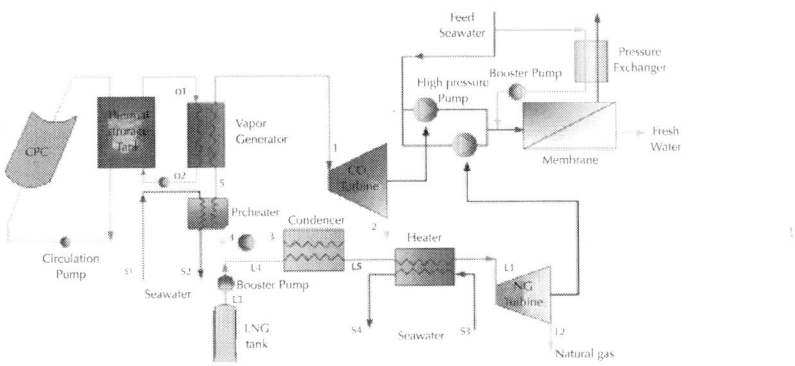

Figure 1: Schematic diagram of the solar-powered transcritical CO_2 power cycle for reverse osmosis desalination based on the recovery of cryogenic energy of LNG.

The solar collectors are used as a main heat source to supply energy to the system. Compound parabolic collector is chosen to collect the solar radiation considering its high collection temperature and intermittent sun-tracking. Thermal storage tank with thermal oil as the working fluid is used to improve the stability and sustainability of the overall system when the solar radiation is not sufficient. The thermal oil is heated to high temperature in the solar collector first, and then flows through the thermal storage tank to store the solar energy. After that, the heat energy is transferred from thermal oil to CO_2 in vapor generator.

The transcritical CO_2 power cycle consists of a booster pump, a vapor generator, a CO_2 turbine and a condenser. CO_2 is compressed to the supercritical state in the booster pump. The supercritical CO_2 enters the vapor generator after preheated in preheater by sea water. In the vapor generator the CO_2 temperature increases by absorbing heat from the thermal oil. The high-temperature supercritical CO_2 enters the CO_2 turbine, where it expands across CO_2 turbine to produce mechanical power to drive a RO high pressure pump. The exhaust CO_2 leaving the CO_2 turbine is condensed in the condenser by LNG.

The LNG subsystem consists of a LNG tank, a LNG pump, a Heater and a NG turbine. LNG with low temperature of −161 °C is removed from the LNG tank and pumped to the required pressure. Then LNG as the heat sink of the transcritical CO_2 power cycle enters the condenser and is vaporized to form natural gas through absorbing heat of CO_2 turbine exhaust and releases the cryogenic energy to condense CO_2 to liquid. The natural gas is delivered to heater to increase its temperature by absorbing the heat from sea water. Then, the natural gas enters LNG turbine to produce the mechanical work to drive another RO high pressure pump.

The RO desalination subsystem consists of three components: high pressure pumps, pressure exchanger and reserve osmosis membranes. The high pressure pumps are driven by mechanical power from the CO_2 turbine and the NG turbine. Sea water containing lots of salt is compressed through pumps. High pressure sea water flows through the membranes and splits into two streams:

fresh water flow and brine flow. High pressure brine flows through the pressure exchanger and delivers pressure energy to feed sea water.

SYSTEM MODELING AND PERFOR-MANCE CRITERION

The Solar Collector Subsystem

The solar collector subsystem consists of the CPC (compound parabolic collector) and the thermal storage tank. The compound parabolic collectors can absorb both beam and diffuse radiation due to its large acceptance angle.

The total effective flux absorbed by CPC is given by [27]

$$S = \left[I_{bm}R_{bm} + \frac{I_{dif}}{C} \right] \tau \rho \alpha \tag{1}$$

Where $I_{bm}R_{bm}$ is the beam radiation flux falling on the aperture plane, and I_{dif}/C is the diffuse radiation flux within the acceptance angle.

The useful heat gain rate Q_u can be calculated by the following equations

$$Q_u = F_R WL \left[S - \frac{U_{lo}}{C} (T_{stg} - T_a) \right] \tag{2}$$

Where

$$F_R = \frac{\dot{m}_{solar}C_{p.oil}}{bU_{lo}L} \left\{ 1 - \exp \left[\frac{F'bU_{lo}L}{\dot{m}_{solar}C_{p.oil}} \right] \right\} \tag{3}$$

Where

$$\frac{1}{F'} = U_{lo} \left[\frac{1}{U_{lo}} + \frac{b}{N\pi Dk} \right] \tag{4}$$

The solar radiation varies greatly with time, reaching the peak in midday and dropping sharply toward zero when the sun sets. The thermal storage tank is an insulated thermal oil storage tank acting as a buffer between the solar collector and the transcritical CO_2 power cycle. It enables the thermal storage temperature to vary in a small range and store the surplus energy of the solar radiation. In order to simplify the calculation of thermal storage tank, the material in thermal storage tank is assumed to be well-mixed to make the storage tank temperature T_{stg} change only with time. The following equation can describe the energy balance in the thermal storage tank.

$$\left[(\rho V c_p)_w + (\rho V c_p)_t\right]\frac{dT_{stg}}{dt} = Q_u - Q_{load} - (UA)_t(T_{stg} - T_a) \tag{5}$$

where (UA) t is the product of the overall heat transfer coefficient and the surface area of thermal storage tank, T_a is the environment temperature around thermal storage tank, $(Vc_p)_w$ is the heat capacity of the oil in the thermal storage tank and $(Vc_p)_t$ is the heat capacity of tank material, Q_{load} is the energy discharged to the transcritical CO_2 power cycle which can be expressed as

$$Q_{load} = \dot{m}_{oil}c_{p,oil}(T_{stg} - T_{O2}) \tag{6}$$

Where the T_{O2} is the temperature of oil from the vapor generator to the thermal storage tank, as shown in Fig. 1.

Transcritical CO_2 Power Cycle Subsystem

The following assumptions are made to simplify the simulation of the transcritical CO_2 power cycle subsystem.

- The pressure drop through the vapor generator and condenser can be negligible.
- Heat transfer with the environment in the subsystem is neglected.
- The turbine and pump respectively have a given isentropic efficiency and mechanical conversion efficiency.

- The stream at the condenser outlet is the saturated liquid.
- The turbine inlet temperature is calculated by giving a constant terminal temperature difference between the vapor generator inlet from the thermal storage tank and the CO_2 turbire inlet.

Fig. 2 shows the T–s diagram of the solar-powered transcritical CO_2 power cycle.

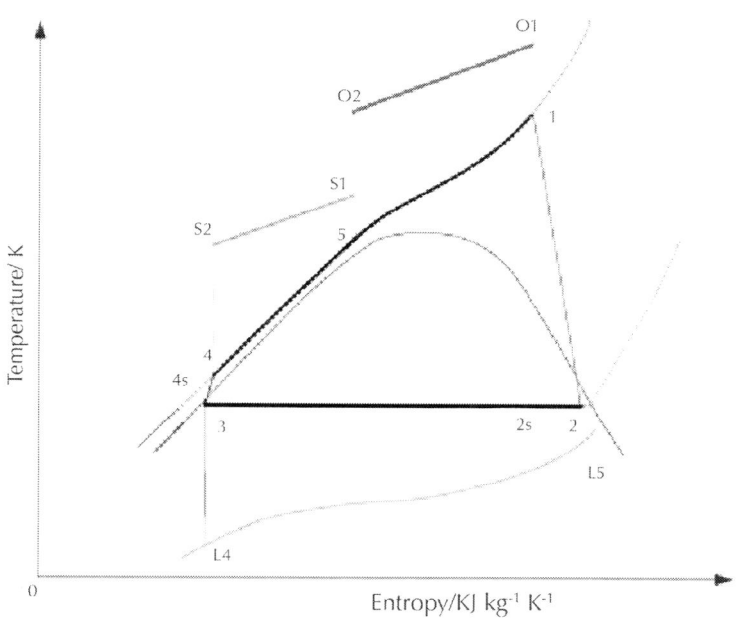

Figure 2: T–s diagram of the solar-powered transcritical CO_2 power cycle.

The heat gained by CO_2 in the vapor generator can be described by the following equations

$$Q_{VG} = \dot{m}_{CO_2}(h_1 - h_5) = Q_{load} \qquad (7)$$

The heat gained from the sea water in the preheater can be calculated by

$$Q_{PH} = \dot{m}_{CO_2}(h_5 - h_4) \qquad (8)$$

The mechanical power produced by CO_2 turbine is expressed as

$$W_{turb,CO_2} = \dot{m}_{CO_2}(h_1 - h_2) \tag{9}$$

The isentropic efficiency of the turbine is

$$\eta_{turb} = \frac{h_1 - h_2}{h_1 - h_{2s}} \tag{10}$$

The pump power consumption is given by

$$W_{pump,CO_2} = \dot{m}_{CO_2}(h_4 - h_3) \tag{11}$$

The isentropic efficiency of pump is

$$\eta_{pump} = \frac{h_{4s} - h_3}{h_4 - h_3} \tag{12}$$

Thus the net power output can be calculated by

$$W_{net,CO_2} = W_{turb,CO_2} - W_{pump,CO_2} \tag{13}$$

The heat transfer in the condenser is expressed as

$$Q_{cnd} = \dot{m}_{CO_2}(h_2 - h_3) = \dot{m}_{LNG}(h_{L5} - h_{L4}) \tag{14}$$

LNG Subsystem

The heat gained from the sea water in the heater is calculated by

$$Q_{HT} = \dot{m}_{LNG}(h_{L1} - h_{L5}) = Q_{cnd} \tag{15}$$

The mechanical power output of the NG turbine can be described as

$$W_{turb,NG} = \dot{m}_{LNG}(h_{L1} - h_{L2}) \tag{16}$$

The pump power consumption in the LNG subsystem is

$$W_{pump,LNG} = \dot{m}_{LNG}(h_{L4} - h_{L3}) \tag{17}$$

Thus the net power output by the LNG subsystem is expressed as

$$W_{net,LNG} = W_{turb,NG} - W_{pump,LNG} \quad (18)$$

Reserve Osmosis Desalination Subsystem

Reverse osmosis based on the principle of selective permeability becomes a modern process technology to produce fresh water. Reverse osmosis occurs when the water is moved across the membrane against the concentration gradient, from lower concentration to higher concentration. In reverse osmosis, pressure is exerted on the side with the concentrated solution to force the water molecules across the membrane to the fresh water side.

The mathematical model for the RO unit is developed by Dessouky [28].

The mass and salt balances are given by

$$\dot{m}_f = \dot{m}_p + \dot{m}_b \quad (19)$$

$$x_f \dot{m}_f = x_p \dot{m}_p + x_b \dot{m}_b \quad (20)$$

where $\dot{M}_f$, $\dot{M}_p$, $\dot{M}_b$ are respectively the feed flow rate, the permeate flow rate, the brine flow rate and x_f, x_p, x_b are the feed salinity, the permeate salinity, the brine salinity.

The feed water mass flow rate $\dot{M}_f$ based on the recovery ratio RR, whose value we chosen is 0.3, and the fresh water mass flow rate $\dot{M}_p$ is

$$\dot{m}_f = \frac{\dot{m}_p}{R_R} \quad (21)$$

The mass flow rate of water passage through a semipermeable membrane is given by:

$$\dot{m}_p = (\Delta p - \Delta \pi)K_w A_m \tag{22}$$

where A_m is the area of reserve osmosis membrane, Kw is the water permeability coefficient, which is expressed as

$$K_w = \frac{6.84 \times 10^{-8} \times [18.68 - (0.177 \times x_b)]}{T_f} \tag{23}$$

where T_f is the temperature of feed water.

Δp is the permeate hydraulic and $\Delta \pi$ is the osmotic pressure, and they can be expressed by

$$\Delta p = \overline{p} - p_p \tag{24}$$

$$\Delta \pi = \overline{\pi} - \pi_p \tag{25}$$

where p_p and π_p are the hydraulic and osmotic pressures of the permeate stream, respectively. $\overline{P}$ And $\overline{\pi}$ are the average feed water pressure and average osmotic pressures on the feed side and brine side, which are given by

$$\overline{p} = 0.5\left(p_f + p_b\right) \tag{26}$$

$$\overline{\pi} = 0.5\left(\pi_f + \pi_b\right) \tag{27}$$

Where p_f and p_b are the hydraulic pressure of feed stream and reject stream. ϖ_f and ϖ_b is the osmotic pressure of feed stream and reject stream.

Osmotic pressures can be expressed by

$$\pi_f = RTx_f \tag{28}$$

$$\pi_b = RTx_b \tag{29}$$

$$\pi_p = RTx_p \tag{30}$$

Where R is the universal gas constant and T is the water temperature.

The mass flow rate of fresh water output driven by the mechanical power W_{net} is estimated as

$$W_{net} = \frac{\Delta P_{net} \dot{m}_f}{\rho_f \eta_{pump}} = W_{net,CO_2} + W_{net,LNG} \tag{31}$$

Where ρ_f is the feed flow rate density, and η_{pump} is the driving pump mechanical efficiency, ΔP_{net} is the net pressure difference across the high pressure pump.

Considering the variation of solar radiation over a day which leads to the change of fresh water output, it is reasonable to calculate the overall output of fresh water to measure the whole day performance by integrating the output over a day. The daily fresh water output can be calculated as follow by integrating the mass flow rate of fresh water

$$V_{day} = \int \frac{\dot{m}_p}{\rho_{water}} dt \tag{32}$$

Exergy Analysis and Performance Criteria

Exergy is the maximum theoretical work which can be obtained from a given form of energy using the environmental parameters as the reference state. Exergy analysis is usually employed to measure the departure of the state of the system from that of the environment. Once the environment is specified, the exergy can be calculated in terms of the property value of the working fluid in the system.

The exergy at a state point i can be defined as

$$E_i = \dot{m}_i[(h_i - h_a) - T_a(s_i - s_a)] \tag{33}$$

The exergy input of the system consists of the exergy of solar radiation and the exergy of LNG. The exergy received by CPC from sun is calculated by

$$E_{solar} = A_{SC}S\left[1 + \frac{1}{3}\left(\frac{T_a}{T_{solar}}\right)^4 - \frac{4}{3}\frac{T_a}{T_{solar}}\right] \tag{34}$$

Where A_{SC} is the solar collector area and S is the global radiation, and T_{solar} is the solar radiation temperature of 6000 K [29].

The exergy input of LNG is

$$E_{LNG} = \dot{m}_{LNG}[(h_3 - h_0) - T_0(s_3 - s_0)] \tag{35}$$

The exergy loss in the solar collector subsystem can be described as

$$I_{SC} = E_{solar} - (E_{O1} - E_{O2}) \tag{36}$$

The exergy loss in the vapor generator is

$$I_{VG} = E_{O1} - E_{O2} - (E_1 - E_5) \tag{37}$$

The exergy loss in the preheater can be calculated as

$$I_{HP} = E_{S1} - E_{S2} - (E_5 - E_4) \tag{38}$$

The exergy loss in condenser is

$$I_{cnd} = E_2 - E_3 - (E_{L5} - E_{L4}) \tag{39}$$

The exergy loss in the CO_2 turbine is

$$I_{turb,CO_2} = E_1 - E_2 - W_{turb,CO_2} \tag{40}$$

The exergy losses in the pump of transcritical CO_2 power cycle subsystem is

$$I_{pump,CO_2} = W_{pump,CO_2} - (E_4 - E_3) \tag{41}$$

The exergy loss in the heater of transcritical CO_2 power cycle subsystem can be calculated as

$$I_{HT} = E_{S3} - E_{S4} - (E_{L5} - E_{L4}) \tag{42}$$

The exergy loss in the pump and turbine of LNG subsystem can be expressed as

$$I_{pump,LNG} = W_{pump,LNG} - (E_{L4} - E_{L3}) \tag{43}$$

$$I_{turb,LNG} = E_{L1} - E_{L2} - W_{turb,LNG} \tag{44}$$

The exergy loss of remain LNG can be expressed as

$$I_{loss} = E_{L2} \tag{45}$$

The overall exergy losses in the system is calculated as

$$I_{sys} = \sum I \tag{46}$$

Exergy efficiency, also called second-law efficiency, is the proper measure of the effectiveness of energy utilization. It can be used to evaluate the performance of the system, which can be defined as

$$\eta_{exg} = \frac{W_{net}}{E_{solar} + E_{LNG}} = \frac{W_{net,CO_2} + W_{net,LNG}}{E_{solar} + E_{LNG}} \tag{47}$$

However, when evaluated the system performance over a whole day, the defined efficiency above suffers from intrinsic limitations. Considering the variation of solar radiation over a day, it is reasonable to define a daily exergy efficiency to demonstrate the whole day performance by integrating the net power output and instantaneous exergy input over a day, which is expressed as follows

$$\eta_{sys,exg} = \frac{W_{day}}{E_{day}} = \frac{\int W_{net} dt}{\int (E_{solar} + E_{LNG}) dt} \tag{48}$$

RESULT AND DISCUSSION

Qingdao in China is selected as the case city to conduct the numerical simulation of the solar-powered transcritical CO_2 power cycle for RO desalination based on the recovery of cryogenic energy of LNG using the mathematical model established above. June 1st is chosen as a typical date assuming it is a sunny day during which the system can experience a periodical variation of the solar radiation. The system simulation is carried out using a simulation program written by authors in Matlab environment. The thermodynamics properties of working fluids are calculated by REFPROP 8.01 [30] developed by the National Institute of Standards and Technology of the United States. Table 1 lists the simulation condition of the system.

Table 1: The simulation condition of the system

City	Qingdao, China
Ambient temperature (K)	298.15
Ambient pressure (MPa)	0.101
Warm sea water temperature (K)	298.15
Area of the aperture of CPC (m²)	235
Number of CPC	42
Specific heat capacity of thermal oil (J kg^{-1} K^{-1})	2350
Concentration ratio of collector	6.50
Reflectivity of concentrator	0.87
Transmissivity of glass cover	0.89
Absorptivity of absorber surface	0.94
Overall loss coefficient of CPC (W m^{-2} K^{-1})	7.50
Heat transfer coefficient on side of absorber tube (W m^{-2} K^{-1})	230
Mass flow rate of oil in CPC (kg s^{-1})	2
Mass flow rate of oil in vapor generator (kg s^{-1})	8
Turbine isentropic efficiency (%)	75
Turbine mechanical efficiency (%)	90
CO_2 turbine inlet pressure (MPa)	14.000
Condenser temperature (K)	233.00
Mass flow rate of LNG (kg s^{-1})	8
NG turbine inlet pressure (MPa)	6.500
NG turbine inlet temperature (K)	290
NG turbine outlet pressure in LNG subsystem (MPa)	4.500
Salinity of feed sea water (kg m^{-3})	45
Salinity of brine water (kg m^{-3})	69
Salinity of fresh water (kg m^{-3})	0.145

Pressure of feed water (MPa)	8.000
Pressure of brine water (MPa)	7.800
Pressure of fresh water (MPa)	0.101

Exergy Analysis

A constant ambient temperature of 298.15 K with the pressure of 0.1 MPa is set as the reference state for exergy calculation despite the ambient temperature varies within a day. Since the exergy input varies with solar radiation over a day, the exergy analysis has been based on the exergy calculation of a day. Table 2shows the daily exergy input, exergy output, and exergy loss of the system on June 1st. It can be seen that the exergy input contains both exergy of solar radiation and exergy of LNG. The losses of LNG and the solar collector subsystem account for the mainly exergy losses in the system. The exergy loss of LNG results from the large amount of LNG exergy input and the low NG turbine outlet temperature. The exergy loss of solar collector subsystem is mainly caused by the low efficiency of CPC. The losses of the vapor generator, preheater, condenser and the heater are mainly result from the temperature difference between the hot fluid and cold fluid. What's more, the larger in the temperature difference, the larger the exergy loss is. Thus, the uses of the preheater and heater could reduce the temperature difference during the heat transfer process, resulting in a reduction in the exergy loss. The other exergy losses which mainly exist in the pump and turbine are related to the irreversibility of process. The exergy output includes the mechanical work produced by transcritical CO_2 power cycle subsystem and the mechanical work produced by LNG subsystem. The mechanical work output of transcritical CO_2 power cycle subsystem mainly contributes to the useful exergy output. The useful exergy output is used to drive the RO desalination subsystem to produce fresh water.

Table 2: The daily exergy input, output, and loss of the system

Term		Value (kWh)	Percentage (%)
Exergy input	Solar collector	46014.82	24.08
	LNG	145092.99	75.92
Exergy losses	CPC and thermal storage unit	28333.50	14.83
	Vapor generator	14134.53	7.40
	Preheater	1905.32	1.00
	CO_2 turbine	3985.04	2.09
	CO_2 Pump	33.38	0.02
	Condenser	5863.25	3.07
	LNG Pump	949.08	0.50
	Heater	10586.23	5.54
	NG turbine	5075.86	2.66
	Loss	111163.49	58.17
Exergy output	CO_2	7077.26	3.70
	LNG	2000.88	1.05

Sensitive Analysis of Key Parameters on the System Performance

The key parameters, namely, CO_2 turbine inlet pressure (P_1), mass flow rate of oil in vapor generator ($\dot{M}_{oil}$), condenser temperature (T_3) and NG turbine inlet pressure (P_{L1}) are examined to evaluate their effects on the performance of the system, namely, the daily net mechanical work output, the daily fresh water output and the daily exergy efficiency. When one parameter varies, the other parameters are kept constant listed in Table 1.

Fig. 3 indicates that the solar radiation and energy discharged to the transcritical CO_2 power cycle from the thermal storage tank on 1st June. It can be seen that the solar radiation begins on about 5:00 and ends on about 19:00 and it reaches its maximal value at noon. The energy discharged to the transcritical CO_2 power cycle has a small variation within a day. In the morning, the solar radiation begins to increase which means the energy input to the thermal storage tank increases. When the solar radiation is strong enough, the energy input to the thermal storage tank is more than the energy discharged to the transcritical CO_2 power cycle, which enable the thermal storage tank temperature to increase, resulting in an increase in the energy discharged to transcritical CO_2 power cycle. In the afternoon, the solar radiation begins to decrease. When the solar radiation is weak enough to enable the energy input to the thermal storage tank to be less than the energy discharged, the thermal storage tank temperature begins to decrease and the energy discharged to transcritical CO_2 power cycle begins to decrease. At night, there is no radiation and the energy discharged to transcritical CO_2 power cycle decreases through all night. Thanks to the existence of thermal storage tank, there is a small variation in the energy discharged to transcritical CO_2 power cycle.

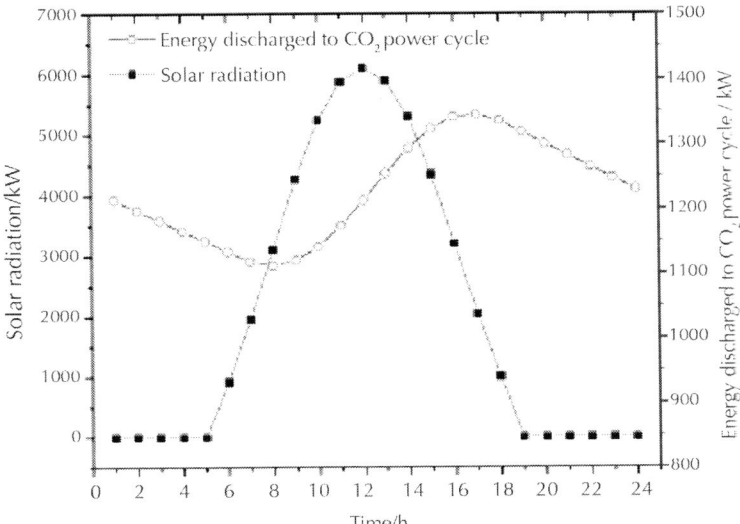

Figure.3: Solar radiation and the energy discharged to transcritical CO_2 power cycle on 1st of June.

Fig. 4 shows the daily variation of the CO_2 turbine inlet temperature and the thermal storage tank temperature as the CO_2 turbine inlet pressure increases. It can be seen that the CO_2 turbine inlet temperature has a small variation over a day and it changes little with an increase in the CO_2 turbine inlet pressure. The thermal storage tank temperature has the same variation with CO_2 turbine inlet temperature. In this system, the CO_2 turbine inlet temperature is determined by the thermal storage temperature which is mainly influenced by the heat absorbed by the CO_2 in vapor generator and the energy input to the thermal storage tank from the CPC. When CO_2 turbine inlet pressure increases, the heat absorbed by CO_2 has almost no change, thus, the thermal storage temperature has little change considering that the daily solar energy input is constant. As a result, the CO_2 turbine inlet temperature is insensitive to the CO_2 turbine inlet pressure. As the CO_2 turbine inlet pressure increases, the mass flow rate of CO_2 increases as shown inFig. 5. The vapor generator outlet enthalpy of CO_2 decreases as the CO_2 turbine inlet pressure increases. Thus the mass flow rate of CO_2 could increase because of the approximately constant heat absorbed by CO_2 from the thermal oil.

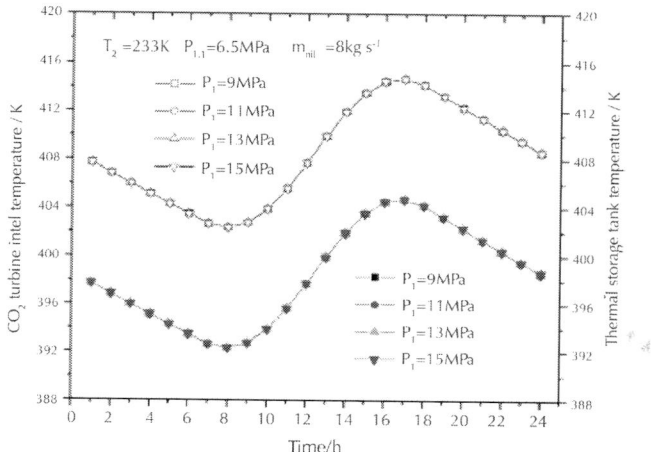

Figure 4: Daily variation of CO_2 turbine inlet temperature and thermal storage tank temperature versus CO_2 turbine inlet pressure.

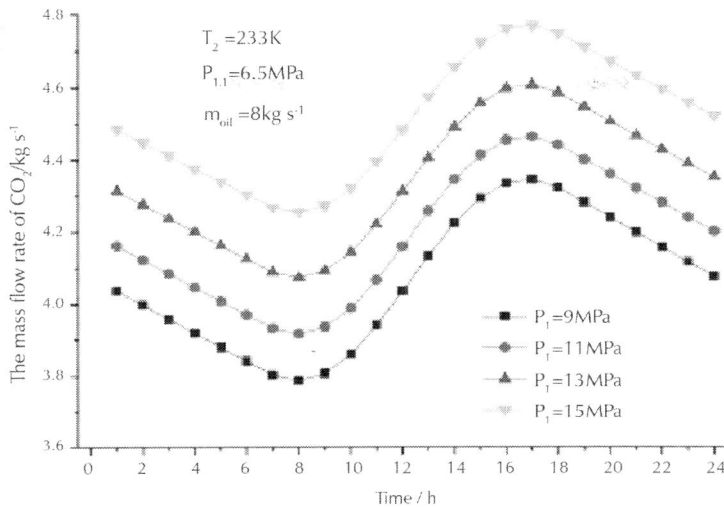

$T_2 = 233K$

$P_{11} = 6.5MPa$

$m_{oil} = 8kg \, s^{-1}$

$P_1 = 9MPa$
$P_1 = 11MPa$
$P_1 = 13MPa$
$P_1 = 15MPa$

The mass flow rate of CO_2/kg s^{-1}

Time / h

Figure 5: Daily variation of mass flow rate of CO_2 versus CO_2 turbine inlet pressure.

Fig. 6 shows the effect of CO_2 turbine inlet pressure on the performance of the system. It can be seen that the daily net mechanical work output first increases and then declines as the CO_2 turbine inlet pressure increases. The peak value approximately turns up at the pressure of 13 MPa. The variation of daily fresh water output is consistent with the daily net mechanical work output. Due to a constant mass flow rate of LNG and the unchanged daily solar exergy input, the daily exergy input is constant. Thus the daily exergy efficiency also varies directly with the daily net mechanical work output. The daily net mechanical work output contains the mechanical work produced by both transcritical CO_2 power cycle subsystem and LNG subsystem. Since the mass flow rate of NG is constant and the parameters in the LNG subsystem has no change when the CO_2 turbine inlet pressure increases, the net mechanical work produced by LNG subsystem keeps constant. It is obvious that the net mechanical work produced by transcritical CO_2 power cycle subsystem is influenced by the mass flow rate of CO_2 and the net CO_2 specific enthalpy drop, the specific enthalpy across the turbine minus the enthalpy across the pump. When the CO_2

specific turbine inlet pressure raises, the CO_2 specific enthalpy drop across the CO_2 turbine increases, the specific enthalpy drop across the pump increases as well, thus the net CO_2 specific enthalpy drop decreases. The mass flow rate of CO_2 would increase with the increase in the CO_2 turbine inlet pressure, resulting in an optimal CO_2 turbine inlet pressure to achieve the maximum daily net mechanical work output.

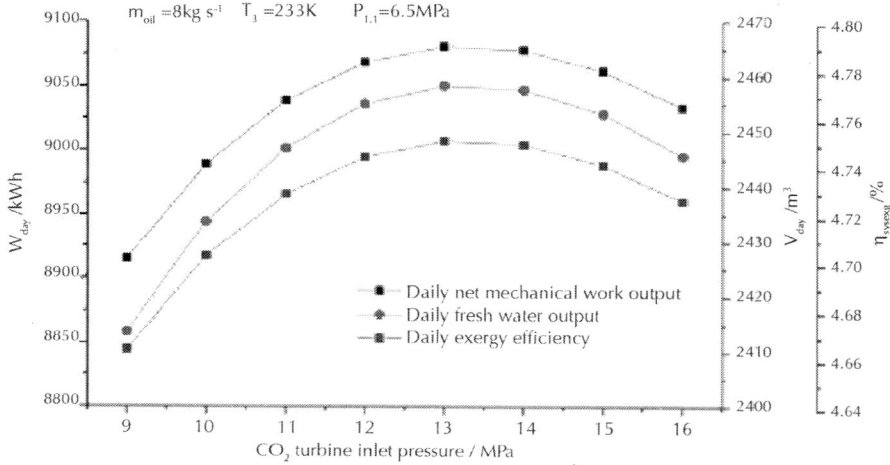

Figure 6: Effects of CO_2 turbine inlet pressure on the system performance.

Fig. 7 shows the daily variation of the CO_2 turbine inlet temperature as the condenser temperature increases. It can be seen that the condenser temperature has no influence on the CO_2 turbine inlet temperature. Due to the use of sea water preheater, the CO_2 temperature at vapor generator inlet keeps constant, resulting in the same daily variation of the CO_2 turbine inlet temperature and no change of the heat absorbed by the CO_2 in the vapor generator. Thus, the condenser temperature has also no effect on the mass flow rate of CO_2 as shown in Fig. 8.

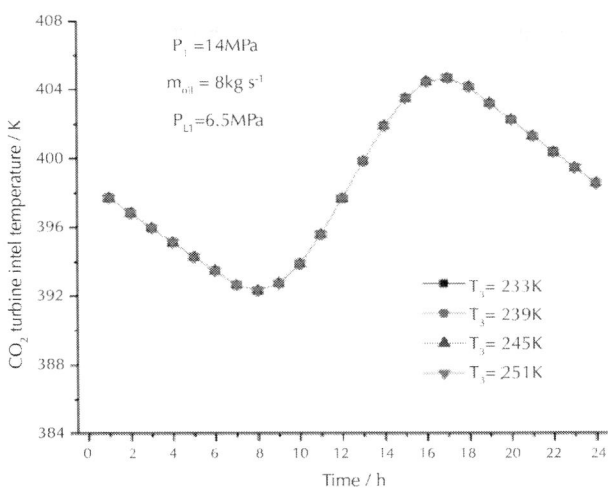

Figure 7: Daily variation of CO_2 turbine inlet temperature versus condenser temperature.

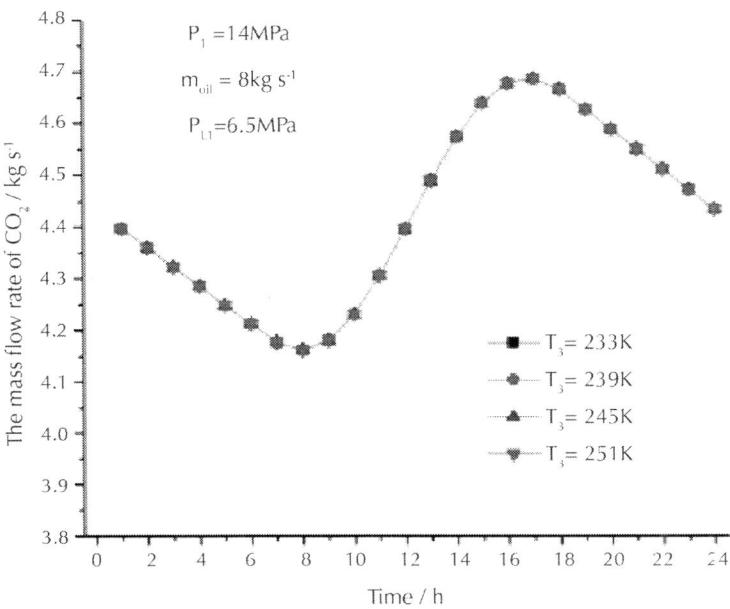

Figure 8: Daily variation of mass flow rate of CO_2 versus condenser temperature.

Fig. 9 shows the effect of condenser temperature on the system performance. It is obvious that the daily net mechanical work output, the daily fresh water output and the daily exergy efficiency all decrease as the condenser temperature increases. The enthalpy drop across the CO_2 turbine decreases when the condenser temperature increases. Thus, the net mechanical work produced in the transcritical CO_2 power cycle subsystem decreases as the condenser temperature increases. The condenser temperature cannot change the NG turbine inlet temperature and pressure considering the sea water heater added in the LNG subsystem. The net mechanical work produced by LNG subsystem keeps constant due the constant mass flow rate of LNG. So the daily net mechanical work output, the daily fresh water output and the daily exergy efficiency decline with an increase in condenser temperature. What's more, it can be seen that the effect of condenser temperature on the system performance is more sensitive than turbine inlet temperature.

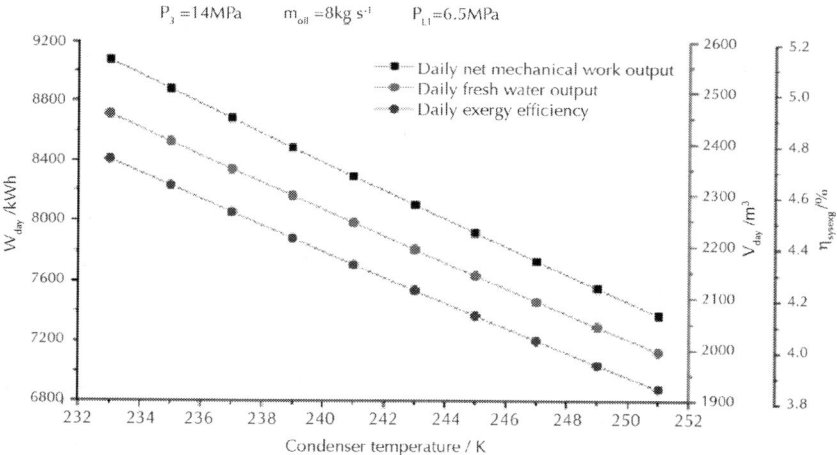

Figure 9: Effects of condenser temperature on the system performance.

Fig. 10 shows the daily variation of CO_2 turbine inlet temperature as the mass flow rate of oil increase. It is found that the CO_2 turbine inlet temperature decreases with an increase in the mass flow rate of oil. An increase in the mass flow rate of oil means a decrease in the temperature difference of oil through the thermal storage tank, which reduce thermal storage tank temperature since the energy input from the CPC is constant. As a result, the CO_2 turbine inlet temperature declines as the thermal storage temperature declines.

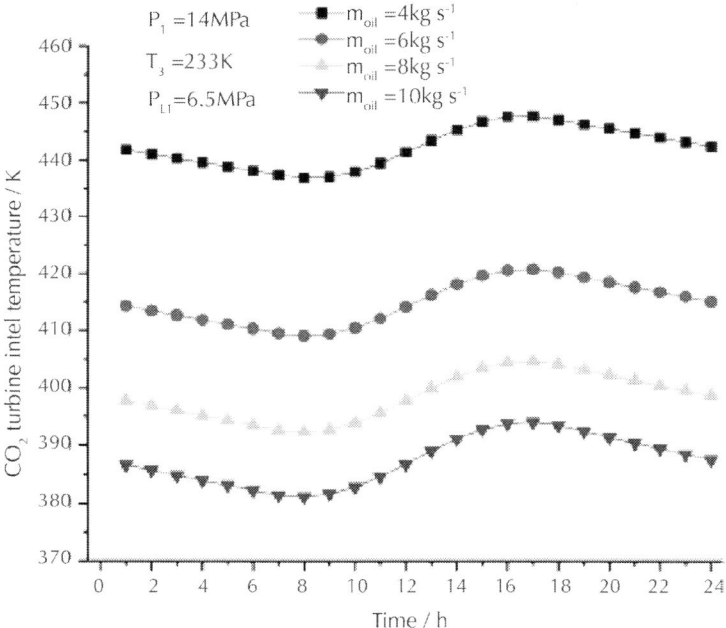

Figure 10: Daily variation of CO_2 turbine inlet temperature versus mass flow rate of oil.

Fig. 11 shows that the mass flow rate of CO_2 increases with an increase in the mass flow rate of oil. The energy output from the thermal storage tank, namely the heat absorbed by CO_2 in the vapor generator, increases as the mass flow rate of oil increases. In addition, the decreased CO_2 turbine inlet temperature leads to a decreased enthalpy difference in the vapor generator. The increased heat absorbed by CO_2 and the decreased enthalpy difference result in an increase in mass flow rate of CO_2.

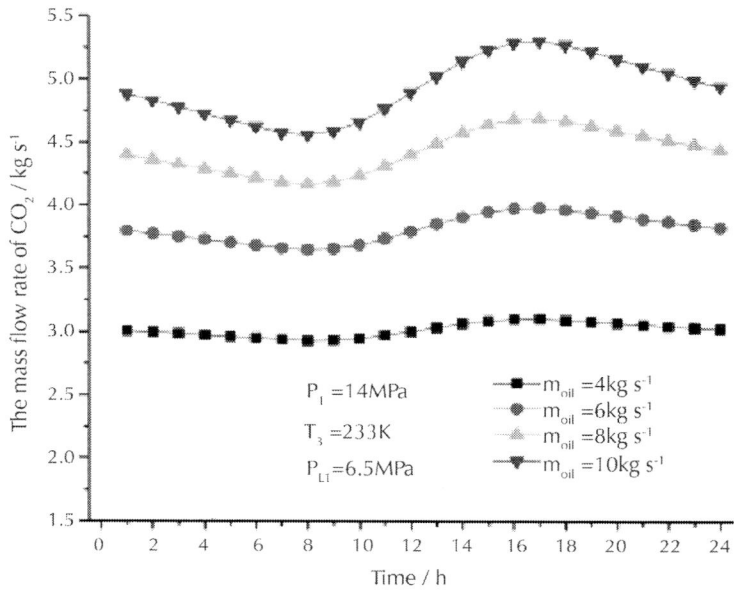

Figure 11: Daily variation of mass flow rate of CO_2 versus mass flow rate of oil.

Fig. 12 indicates that the daily net mechanical work output increases as the mass flow rate of oil increases. The daily fresh water output and daily exergy efficiency have the same variation as the daily net mechanical work output. Although the decreased CO_2 turbine inlet temperature could lead to a decrease in enthalpy drop across the CO_2 turbine, the increased mass flow rate of CO_2 enables the net mechanical work output of transcritical CO_2 power cycle subsystem to increase. Due to the unchanged net mechanical work output of LNG subsystem, the daily net mechanical work output increases, resulting in an increase in the daily fresh water output and daily exergy efficiency.

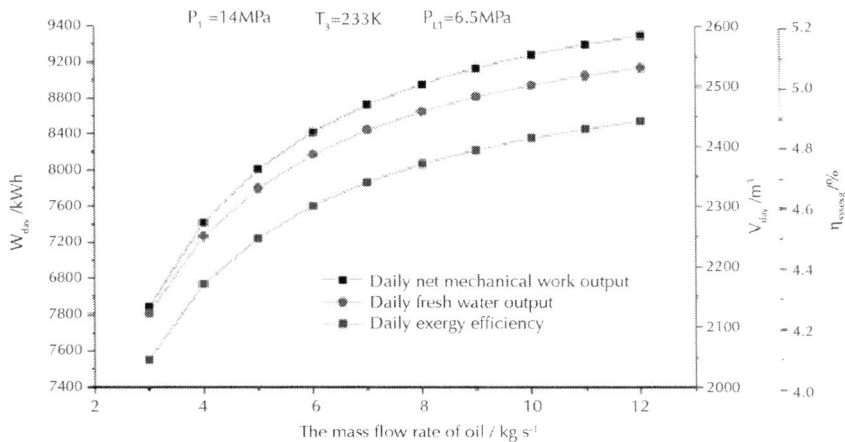

Figure 12: Effects of mass flow rate of oil on the system performance.

Fig. 13 shows the effects of NG turbine inlet pressure on the system performance. It can be seen that the daily net mechanical work output, the daily fresh water output and the daily exergy efficiency increases with the raise of NG turbine inlet pressure. The NG turbine inlet pressure mainly influences the net mechanical work output of LNG subsystem. The nature gas with a high pressure leading to the high pressure ratio causes an increased enthalpy drop across the NG turbine, resulting in an increase in mechanical work of NG turbine. Due to unchanged net mechanical work in the transcritical CO_2 power cycle subsystem, the daily net mechanical work in the LNG subsystem increases, resulting in an increase in daily net mechanical work output of overall system. Accordingly, the daily exergy efficiency and the daily fresh water output both increases with an increase in NG turbine inlet pressure. In addition, a high pressure of LNG can reduce the exergy loss in the condenser due to a decreased temperature difference in the condenser, contributing to an increase in exergy efficiency.

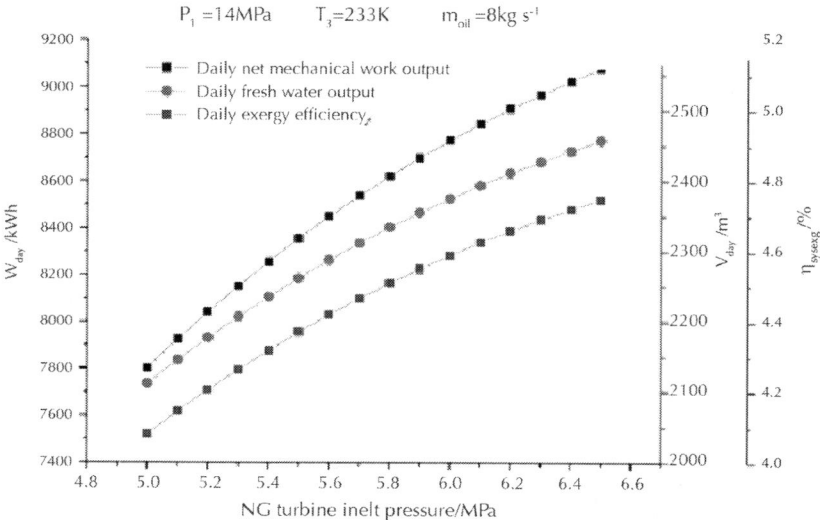

Figure 13: Effects of LNG subsystem turbine inlet pressure on the system performance.

System Optimization

By sensitive analysis of key system parameters, it is found that the key parameters have significant effects on the system performance and the system performance is determined by the combination of the key system parameters. Thus, in order to obtain the best performance of the system, it is essential to conduct the parametric optimization to find the optimal value of each key parameter. In this study, GA (genetic algorithm) [31] as one of the effective intelligent optimization methods is employed to achieve the parametric optimization and obtain the optimum combination of the key parameters. GA is an evolutionary algorithm inspired by evolutionary biology that uses some techniques such as selection, crossover and mutation to search for the best solution. GA, as a powerful and broadly applicable stochastic search and optimization technique, is perhaps the most widely known types of evolutionary computation method today [32]. GA has been applied many areas such as in science, industrial engineering and economy. The procedure of GA can be described as follow. The first step is the initialization of a population of individuals. Each individual means a possible solution to the problem and it is generated randomly depending on the natural of the problem. A fitness function is employed to evaluate the performance of each individual. Some individuals are selected based on its fitness to reproduce new individuals. The reproduction methods include mutation and crossover by means of genetic operations. Mutation is a type of transformations which generates new individuals by making change of a single individual and crossover is a way generating new individuals by combining parts of two individuals. A new population is formed by the new generated individuals. The new population is also to be evaluated to measure its fitness. After several generations, the algorithm could converge to the best individuals.

The performance of the system includes daily net mechanical work output, daily fresh water output and daily exergy efficiency. Considering the daily fresh water output and daily exergy efficiency are consistent with the daily mechanical work output. The maximum daily net mechanical work output is selected as the optimization objective. The CO_2 turbine inlet temperature, the mass flow rate of oil, the condenser temperature and the NG turbine inlet pressure are selected as decision variables. Table 3 presents the condition of the parameter optimization and the optimization results. It can be seen that the optimal turbine inlet pressure exists within the range and the other three parameters, namely, the optimal condenser temperature, the optimal NG turbine inlet pressure and the optimal mass flow rate of oil, are all on the boundary of their range. The four optimal parameters all match the analysis above. The proposed system can reach the maximum daily exergy efficiency of 4.90% and provide 2537.33 m^3 fresh water on the June 1st. What's more, LNG has a significant contribution to the fresh water output not only for its decline in the condenser temperature but also for its mechanical work output.

Table 3: The condition and results of the parameter optimization

Range of CO_2 turbine inlet pressure (MPa)	9.00–16.00
Range of condenser temperature (K)	233.0–251.0
Range of mass flow rate of oil (kg s^{-1})	3.00–12.00
Range of NG turbine inlet pressure (MPa)	5.00–6.50
Optimal CO_2 turbine inlet pressure (MPa)	12.28
Optimal condenser temperature (K)	233
Optimal mass flow rate of oil (kg s^{-1})	12
Optimal NG turbine inlet pressure (MPa)	6.50
Daily net mechanical work output of CO_2 (kWh)	7371.27
Daily net mechanical work output of natural gas (kWh)	2000.88
Daily net mechanical work output (kWh)	9372.15
Daily exergy efficiency (%)	4.90
Daily fresh water output (m^3 d^{-1})	2537.33
Optimal membrane area (m^2)	3500

CONCLUSIONS

In the present study, a solar-powered transcritical CO_2 power cycle for RO desalination based on the recovery of cryogenic energy of LNG is proposed to produce fresh water by renewable energy. The effects of some key parameters on the system performance are examined, and a parametric optimization is conducted to obtain the optimal performance by GA. The main conclusions drawn from the study are listed as follows:

- By introducing a thermal storage tank into the system, the solar-powered transcritical CO_2 power cycle for RO desalination can provide a continuous fresh water production over a long time.
- The losses of LNG and the solar collector subsystem account for the mainly exergy losses in the system.
- The CO_2 turbine inlet pressure has an optimal value to

reach the maximum exergy efficiency of the system. As the condenser temperature increases, the daily exergy efficiency of the system decreases. As the mass flow rate of oil and the NG turbine inlet pressure increase, the daily exergy efficiency of the system increases.

- Through parametric optimization, the system can reach the daily exergy efficiency of 4.90% and provide 2537.33 m^3 fresh water per day under the given conditions.

ACKNOWLEDGMENTS

The authors gratefully acknowledge the financial support by the National Natural Science Foundation of China (Grant No. 51106117).

REFERENCES

1.   Delgado-Torres AM, Garcia-Rodriguez L. Status of solar thermal-driven reverse osmosis desalination. Desalination 2007;216:242e51.

2.   Ghermandi A, Messalem R. Solar-driven desalination with reverse osmosis: the state of the art. Desalin Water Treat 2009;7:285e96.

3.   Delgado-Torres AM, Garcia-Rodriguez L, Romero-Ternero VJ. Preliminary design of a solar thermal-powered seawater reverse osmosis system. Desalination 2007;216:292e305.

4.   Delgado-Torres AM, Garcia-Rodriguez L. Double cascade organic rankine cycle for solar-driven reverse osmosis desalination. Desalination 2007;216:306e13.

5.   Kosmadakis G, Manolakos D, Kyritsis S, Papadakis G. Simulation of an autonomous, two-stage solar organic Rankine cycle system for reverse osmosis desalination. Desalin Water Treat 2009;1:114e27.

6.   Kosmadakis G, Manolakos D, Kyritsis S, Papadakis G. Economic assessment of a two-stage solar organic Rankine cycle for reverse osmosis desalination. Renew Energy 2009;34:1579e86.

7.   Kosmadakis G, Manolakos D, Kyritsis S, Papadakis G. Design of a two stage organic Rankine cycle system for reverse osmosis desalination supplied from a steady thermal source. Desalination 2010;250:323e8.

8.   Kosmadakis G, Manolakos D, Kyritsis S, Papadakis G. Comparative thermodynamic study of refrigerants to select the best for use in the hightemperature stage of a two-stage organic Rankine cycle for RO desalination. Desalination 2009;243:74e94.

9.   Manolakos D, Kosmadakis G, Kyritsis S, Papadakis G. Identification of behaviour and evaluation of performance of small scale, low-temperature organic Rankine cycle system coupled with a RO desalination unit. Energy 2009;34: 767e74.

10. Nafey AS, Sharaf MA. Thermo-economic analysis of a combined solar organic Rankine cycle-reverse osmosis desalination process with different energy recovery configurations. Desalination 2010;261:138e47.

11. Manolakos D, Kosmadakis G, Kyritsis S, Papadakis G. On site experimental evaluation of a low-temperature solar organic Rankine cycle system for RO desalination. Sol Energy 2009;83:646e56.

12. Manolakos D, Papadakis G, Kyritsis S, Bouzianas K. Experimental evaluation of an autonomous low-temperature solar Rankine cycle system for reverse osmosis desalination. Desalination 2007;203:366e74.

13. Kim MH, Pettersen J, Bullard CW. Fundamental process and system design issues in CO_2 vapor compression systems. Prog Energy Combust 2004;30: 119e74.

14. Beckman EJ. Supercritical and near-critical CO2 in green chemical synthesis and processing. J Supercrit Fluid 2004;28(2):121e91.

15. Cayer E, Galanis N, Nesreddine H. Parametric study and optimization of a transcritical power cycle using a low temperature source. Appl Energy 2010;87(4):1349e57.

16. Cayer E, Galanis N, Desilets M, Nesreddine H, Roy P. Analysis of a carbon dioxide transcritical power cycle using a low temperature source. Appl Energy 2009;86(7):1055e63.

17. Zhang XR, Yamaguchi H, Uneno D, Fujima K, Enomoto M, Sawada N. Analysis of a novel solar energy-powered Rankine cycle for combined power and heat generation using supercritical carbon dioxide. Renew Energy 2006;31:1839e54.

18. Zhang XR, Yamaguchi Hiroshi, Yuhui Cao. Hydrogen production from solar energy powered supercritical cycle using carbon dioxide. Int J Hydrogen Energy 2010;35:4925e32.

19. Yamaguchi H, Zhang XR, Fujima K, Enomoto M, Sawada N. Solar energy powered Rankine cycle using supercritical CO2. Appl Therm Eng 2006;26: 2345e54.

20. Niu XD, Yamaguchi H, Iwamoto Y, Zhang XR. Optimal arrangement of the solar collectors of a supercritical CO2-based solar Rankine cycle system. Appl Therm Eng 2013;50:505e10.

21. Zhang XR, Yamaguchi H, Uneno D. Experimental study on the performance of solar Rankine system using supercritical CO2. Renew Energy 2007;32:2617e 28.

22. Niu XD, Yamaguchi H, Zhang XR, Iwamoto Y, Hashitani N. Experimental study of heat transfer characteristics of supercritical CO2 fluid in collectors of solar Rankine cycle system. Appl Therm Eng 2011;31:1279e85.

23. Zhang N, Lior N. A novel near-zero CO2 emission thermal cycle with LNG cryogenic exergy utilization. Energy 2006;31(10, 11):1666e79.

24. Lin W, Huang M, He H. A transcritical CO Rankine cycle with LNG cold energy utilization and liquefaction of CO in gas turbine exhaust. J Energy Resour 2009;131:042201.

25. Song YH, Wang JF, Dai YP, Zhou EM. Thermodynamic analysis of a transcritical CO2 power cycle driven by solar energy with liquefied natural gas as its heat sink. Appl Energy 2012;92:194e203.

26. Sun ZX, Wang JF, Dai YP, Wang JH. Exergy analysis and optimization of a hydrogen production process by a solar-liquefied natural gas hybrid driven transcritical CO2 power cycle. Int J Hydrogen Energy 2012;37:18731e9.

27. Sukhatme SP. Solar energy principles of thermal collection and storage. India: McGraw-Hill Publishing Company Limited; 1984.

28. El-Dessouky Hisham T, Ettouney Hisham M. Fundamentals of salt water desalination. Amsterdam: Elsevier Science B.V; 2002.

29. Banat F, Jwaied N. Exergy analysis of desalination by solar-powered membrane distillation units. Desalination 2008;230:27e40.

30. NIST Standard Reference Database 23. NIST thermodynamic and transport properties of refrigerants and refrigerant mixtures REFPROP. Version 8.01; 2007.

31. Holland J. Adaptation in nature and artificial systems: an introductory analysis with applications to biology, control and artificial intelligence. Massachusetts: MIT Press; 1992.

32. Gen M, Cheng RW. Genetic algorithms and engineering optimization. USA: John Wiley & Sons, Inc; 2000.

Hydrogen Production by Steam Reforming of Liquefied Natural Gas (LNG) Over Mesoporous Ni-Al$_2$O$_3$ Aerogel Catalyst Prepared by a Single-step Epoxide-driven Sol-gel Method

Yongju Bang, Jeong Gil Seo, Min Hye Youn, and In Kyu Song

School of Chemical and Biological Engineering, Institute of Chemical Processes, Seoul National University, Shinlim-dong, Kwanak-ku, Seoul 151-744, South Korea

ABSTRACT

A mesoporous Ni-Al$_2$O$_3$ aerogel catalyst was prepared by a single-step epoxide-driven sol-gel method and a subsequent supercritical CO$_2$ drying method (NA-ES catalyst). For comparison, a mesoporous Ni-Al$_2$O$_3$ aerogel catalyst was also prepared by a single-step alkoxide-based sol-gel method and a subsequent supercritical CO$_2$ drying method (NA-AS catalyst). Differences in physicochemical properties and catalytic activities of mesoporous Ni-Al$_2$O$_3$ aerogel catalysts in the steam reforming of liquefied natural gas (LNG) were investigated. Textural properties of Ni-Al$_2$O$_3$ aerogel catalysts were affected by the preparation method. Nickel species were highly dispersed in alumina through the formation of nickel aluminate phase in both NA-ES and NA-AS catalysts. However, chemical states of Al atoms in both catalysts were quite different. In addition, nickel species in the NA-ES catalyst exhibited high reducibility and high dispersion compared to those in the NA-AS catalyst. In the steam reforming of LNG, NA-ES catalyst exhibited a better catalytic performance than NA-AS catalyst in terms of LNG conversion and hydrogen yield, although both catalysts showed a stable catalytic performance during the reaction without deactivation behavior. Furthermore, NA-ES catalyst with small average nickel diameter suppressed water-gas shift reaction. Reducibility and dispersion of nickel species served as important factors determining the catalytic performance of the catalysts.

INTRODUCTION

Hydrogen has been considered as the most viable energy carrier due to emission problem of fossil fuels[1], [2] and [3]. In particular, technological advances in hydrogen utilization such as fuel cell make hydrogen more important as an alternative and promising fuel [4] and [5]. Although splitting of water through photo-catalysis and electrolysis is known as an ultimate method for producing hydrogen, low productivity and high cost make it unfavorable for commercial hydrogen production [6]. Instead, several catalytic reforming processes for commercial hydrogen production from hydrocarbons have been extensively investigated, including steam reforming [7] and [8], partial oxidation [9] and [10], auto-thermal reforming [11] and [12], and dry reforming [13]. Among these reforming processes, steam reforming has been widely employed for hydrogen production due to its high economical advantage. Liquefied natural gas (LNG), which is abundant and mainly composed of methane, can be used as a promising source for hydrogen production by steam reforming reaction. As LNG pipelines are more widespread in the modern cities, LNG will become a promising hydrogen source for fuel cell system equipped with fuel processing unit.

It has been reported that noble metal-based catalysts showed a superior catalytic performance compared to the other transition metal-based catalysts in the steam reforming reaction [14]. Nickel-based catalysts, however, have been regarded as the most feasible catalyst for steam reforming reactions because of their low cost and high intrinsic activity for C–C and O–H bond cleavage, a though nickel-based catalysts experience a severe catalyst deactivation caused by carbon deposition and nickel sintering dur ng the reactions [15] and [16]. Therefore, many researches have been carried out to solve deactivation problem of nickel-based catalysts [17], [18] and [19]. For example, nickel catalysts supported on various metal oxides have been investigated for steam reforming of methane due to their excellent textural properties which make finely dispersed active nickel species throughout the catalysts and enhance heat/mass transfer over the catalysts [19], [20] and [21]. In particular, it has been reported that nickel-alumina aerogel catalyst prepared by a single-step alkoxide-based sol-gel method and a subsequent CO_2 supercritical drying method is highly active and very stable in the steam reforming reaction [22] and [23].

Traditionally, aerogel materials have been prepared by an alkoxide-based sol-gel method using metal alkoxide precursors such as aluminum tri-sec-butoxide [24]. However, these metal alkoxide precursors are expensive and difficult to handle due to their high reactivity and sensitivity to heat and moisture. To tackle these problems, a new versatile sol-gel synthesis route utilizing simple inorganic salt precursors and epoxide has been proposed with an aim of preparing mesoporous metal oxides such as iron oxide [25], alumina [26], and titania [27]. In this method, epoxide serves as an acid scavenger and gelation agent in the sol-gel polymerization process as presented in Fig. 1. Hydrated metal cations, which act as strong acid due to charge transfer from the coordinated water molecules, protonate oxygen atoms of epoxide molecules. Because this protonation step is very fast, many hydroxyl groups are rapidly generated on the metal cations. The resulting hydroxyl groups on the metal cations then induce polycondensation reaction through rapid formation of metal–O–metal bond. The protonated epoxide molecules are irreversibly ring-opened by nucleophiles in the solution, and finally, they are removed through drying or calcination step.

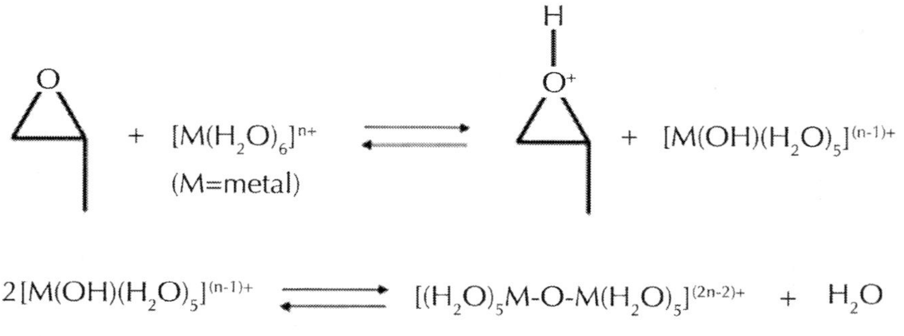

Figure 1: Reactions involved in the preparation of metal oxide by an epoxide-driven sol-gel method.

To our best knowledge, however, a mesoporous Ni-Al$_2$O$_3$ aerogel catalyst prepared by an epoxide-driven sol-gel method has never been employed for hydrogen production by steam reforming of LNG. It is expected that physicochemical properties of Ni-Al$_2$O$_3$ catalyst prepared by an epoxide-driven sol-gel method would be quite different from those of Ni-Al$_2$O$_3$ catalyst prepared by an alkoxide-based sol-gel method due to different gelation mechanism of the catalysts. Therefore, a systematic investigation on the application of mesoporous Ni-Al$_2$O$_3$ aerogel catalyst prepared by a single-step epoxide-driven sol-gel method to hydrogen production by steam reforming of LNG would be worthwhile.

In this work, a mesoporous Ni-Al$_2$O$_3$ aerogel catalyst was prepared by a single-step epoxide-driven sol-gel method and a subsequent supercritical CO$_2$ drying method (NA-ES catalyst). For comparison, a mesoporous Ni-Al$_2$O$_3$ aerogel catalyst was also prepared by a single-step alkoxide-based sol-gel method and a subsequent supercritical CO$_2$ drying method (NA-AS catalyst). The prepared mesoporous Ni-Al$_2$O$_3$ aerogel catalysts were applied to the hydrogen production by steam reforming of LNG. Differences in physicochemical properties and catalytic activities of Ni-Al$_2$O$_3$ aerogel catalysts (NA-ES and NA-AS) in the steam reforming of LNG were investigated.

EXPERIMENTAL

Preparation of Mesoporous Ni-Al$_2$O$_3$ Aerogel Catalysts

A mesoporous Ni-Al$_2$O$_3$ aerogel catalyst was prepared by a single-step epoxide-driven sol-gel method and a subsequent supercritical CO$_2$ drying method [26]. 8.6 g of aluminum precursor (aluminum nitrate nonahydrate, Sigma–Aldrich) and 2.4 g of nickel precursor (nickel nitrate hexahydrate, Sigma–Aldrich) were dissolved in ethanol (90 ml) with vigorous stirring for hydration of Al^{3+} and Ni^{2+} ions. Propylene oxide as a gelation agent was then added into the metal precursor solution to make hydroxyl group on the hydrated ions and to induce polycondensation reaction between Al^{3+} and Ni^{2+} ions. Molar ratio of aluminum precursor:nickel precursor:propylene oxide was fixed at 1:0.35:13.5. After maintaining the resulting solution for several minutes, a green opaque nickel-alumina composite gel was formed. The gel was aged for 2 days, and then it was dried at 50 °C and 120 atm for 18 h in a stream of supercritical

CO_2. The resulting powder was then calcined at 700 °C for 5 h to yield a mesoporous Ni-Al_2O_3 aerogel catalyst. The Ni-Al_2O_3 aerogel catalyst prepared by a single-step epoxide-driven sol-gel method was denoted as NA-ES.

For comparison, a mesoporous Ni-Al_2O_3 aerogel catalyst was prepared by a single-step alkoxide-based sol-gel method and a subsequent supercritical CO_2 drying method, according to the similar methods reported in literatures [22] and [23]. 7.0 g of aluminum precursor (aluminum tri-sec-butoxide, Sigma–Aldrich) was dissolved in ethanol (60 ml) at 80 °C with vigorous stirring. Small amounts of distilled water and nitric acid, which had been diluted with ethanol, were slowly added into the solution of aluminum precursor for partial hydrolysis of aluminum precursor. After maintaining the resulting solution at 80 °C for a few minutes, a clear sol was obtained. The sol was cooled to 60 °C, and 2.4 g of nickel precursor (nickel acetate tetrahydrate, Sigma–Aldrich) was slowly added into the sol to obtain a nickel-alumina composite sol. After cooling the nickel-alumina composite sol to room temperature, a monolithic gel was obtained by adding an appropriate amount of water diluted with ethanol into the sol. The gel was aged for 2 days, and then it was dried at 50 °C and 120 atm for 18 h in a stream of supercritical CO_2. The resulting powder was finally calcined at 700 °C for 5 h to yield a mesoporous Ni-Al_2O_3 aerogel catalyst. The Ni-Al_2O_3 aerogel catalyst prepared by a single-step alkoxide-based sol-gel method was denoted as NA-AS. Nickel loading in the NA-ES and NA-AS catalysts was fixed at 40 wt%.

In addition, mesoporous alumina aerogel supports were prepared by an epoxide-driven sol-gel method and an alkoxide-based sol-gel method without using nickel precursor, according to the similar methods described above. The prepared mesoporous alumina aerogels prepared by an epoxide-driven sol-gel method and an alkoxide-based sol-gel method were denoted as A-ES and A-AS, respectively.

Characterization

Nitrogen adsorption–desorption isotherms of NA-ES and NA-AS catalysts were obtained with an ASAP-2010 (Micromeritics) instrument. Chemical compositions of NA-ES and NA-AS catalysts were determined by ICP-AES (ICPS-1000IV, Shimadzu) analyses. Crystalline structures of the catalysts were investigated by XRD (D-Max2500-PC, Rigaku) measurements using Cu-K radiation (= 1.541 Å) operated at 50 kV and 150 mA. Chemical states of aluminum species in the A-ES and A-AS supports were examined by ^{27}Al MAS NMR (magic angle spinning nuclear magnetic resonance) spectra, which were obtained with a Bruker AVANCE II spectrometer (500 MHz) at an MAS frequency of 11 kHz. In order to examine the interaction between nickel species and alumina, temperature-programmed reduction (TPR) measurements were carried out in a conventional flow system with a moisture trap connected to a thermal conductivity detector (TCD) at temperatures ranging from room temperature to 1000 °C with a ramping rate of 5 °C/min. For the TPR measurements, a mixed stream of H_2 (2 ml/min) and N_2 (20 ml/min) was used for 50 mg of catalyst sample. Hydrogen chemisorption experiments (BELCAT-B, BEL Japan) were conducted to measure the nickel surface area, nickel dispersion, and average nickel diameter of the catalysts. Prior to the chemisorption measurements, 50 mg of each catalyst was reduced with a mixed stream of hydrogen (2.5 ml/min) and argon (47.5 ml/min) at 700 °C for 3 h, and subsequently, it was purged with pure argon (50 ml/min) for 15 min at 700 °C. The sample was then cooled to 50 °C under a flow of argon (50 ml/min). The amount of hydrogen uptake was measured by periodically injecting diluted hydrogen (5% hydrogen and 95% argon) into the reduced catalyst using an on-line sampling valve. Nickel surface area, nickel dispersion, and average nickel diameter were calculated by assuming that one hydrogen atom occupies one surface nickel atom. The amount of carbon deposition in the used catalysts was determined by CHNS elemental analyses (CHNS 932, Leco).

Hydrogen Production by Steam Reforming of LNG

Hydrogen production by steam reforming of LNG was carried out in a continuous flow fixed-bed reactor at 600 °C under atmospheric pressure. Prior to the catalytic reaction, each catalyst (100 mg) was reduced with a mixed stream of H_2 (3 ml/min) and N_2 (30 ml/min) at 700 °C for 3 h. Feed composition was fixed at $CH_4:C_2H_6:H_2O:N_2$ = 4.6:0.4:10:30. Total feed rate with respect to catalyst weight was maintained at 27,000 ml/h g. Reaction products were periodically sampled and analyzed using an on-line gas chromatograph (ACME 6000, Younglin) equipped with a thermal conductivity detector (TCD). LNG conversion, hydrogen yield, and composition of outlet gas were calculated according to the following equations.

$$LNG \; conversion \; (\%) = \left(1 - \frac{F_{CH_4, \, out} + F_{C_2H_6, \, out}}{F_{CH_4, \, in} + F_{C_2H_6, \, in}} \right) \times 100 \tag{1}$$

$$Hydrogen \; yield \; (\%) = \frac{F_{H_2, \, out}}{2 \times F_{CH_4, \, in} + 3 \times F_{C_2H_6, \, in}} \times 100 \tag{2}$$

$$Composition \; of \; species \; i \; in \; outlet \; gas \; (\%) = \frac{F_{i, \, out}}{F_{total, \, out} - F_{N_2, \, out}} \times 100 \tag{3}$$

RESULTS AND DISCUSSION

Textural Properties of Ni-Al$_2$O$_3$ Aerogel Catalysts

Textural properties of NA-ES and NA-AS catalysts were examined by nitrogen adsorption–desorption isotherm measurements as represented in Fig. 2. Both catalysts exhibited IV-type isotherms indicative of capillary condensation in well-developed mesoporous structure [28]. It is noticeable that NA-ES catalyst showed hysteresis loop similar to H1-type, while NA-AS catalyst exhibited H2-type hysteresis loop. It has been reported that IV-type isotherm with H1-type hysteresis loop appears in mesoporous materials comprising nearly spherical-shaped agglomerates or compacts [29]. On the other hand, IV-type isotherm with H2-type hysteresis loop is closely associated with the "ink-bottle" pore structure whose pore shape is not well-defined [26] and [29]. Thus, it can be inferred that NA-ES catalyst consists of fairly regular spherical-shaped particles compared to NA-AS catalyst.

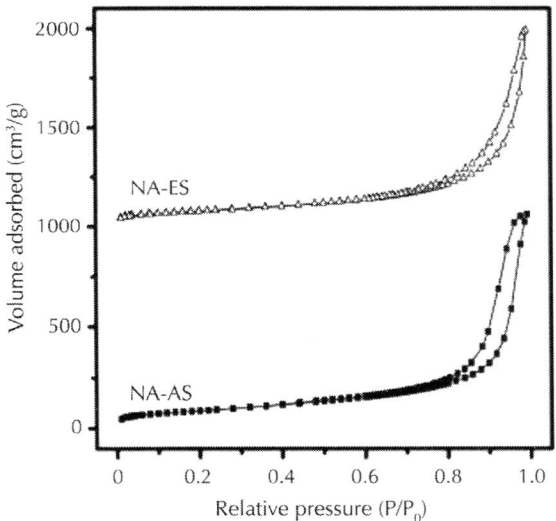

Figure 2: Nitrogen adsorption–desorption isotherms of NA-ES and NA-AS catalysts calcined at 700 °C. The adsorption–desorption data for NA-ES and NA-AS catalysts were vertically offset by 0 and 1000 cm^3/g, respectively.

Detailed textural properties of NA-ES and NA-AS catalysts are summarized in Table 1. Both catalysts showed high surface area, large pore volume, and large average pore diameter, representing successful preparation of mesoporous aerogel materials. It is believed that different pore volumes and pore sizes of these catalysts were due to different pore formation mechanism between epoxide-driven sol-gel method and alkoxide-based sol-gel method.

Table 1: Textural properties of NA-ES and NA-AS catalysts calcined at 700 °C for 5 h

Catalyst	Ni/Al atomic ratio[a]	Surface area (m²/g)[b]	Pore volume (cm³/g)[c]	Average pore diameter (nm)[d]
NA-ES	0.35	282	1.54	15.8
NA-AS	0.35	296	1.60	16.3

[a]Determined by ICP-AES measurement.

[b]Calculated by the BET equation.

[c]BJH desorption pore volume.

[d]BJH desorption average pore diameter.

Crystalline Structure of Calcined Ni-Al₂O₃ Aerogel Catalysts

Fig. 3 shows the XRD patterns of NA-ES and NA-AS catalysts calcined at 700 °C. It was found that NA-ES and NA-AS catalysts exhibited no characteristic diffraction peaks corresponding to bulk nickel oxide (solid lines in Fig. 3). Instead, both NA-ES and NA-AS catalysts showed three distinct diffraction peaks corresponding to nickel aluminate phase (closed circles in Fig. 3). It should be noted that the diffraction peaks of -Al₂O₃ (440) (dashed line in Fig. 3) shifted to a lower diffraction angle in both catalysts. This is because cationic deficient sites in the lattice of Al₂O₃ were expanded by the incorporation of Ni^{2+} ions[30] and [31]. From these results, it can be inferred that nickel species were finely dispersed in alumina as a form of nickel aluminate with the strong interaction between nickel and alumina support derived from Ni–O–Al composite structure.

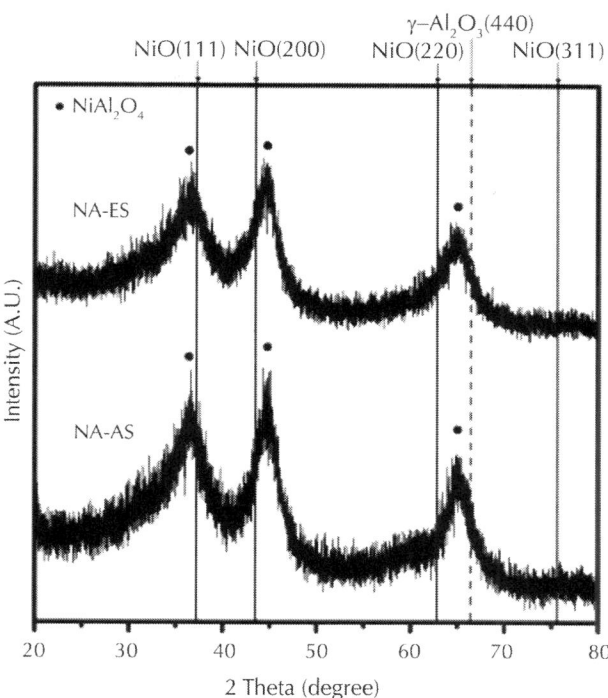

Figure 3: XRD patterns of NA-ES and NA-AS catalysts calcined at 700 °C.

Effect of Preparation Method on Chemical States of Aluminum Species

^{27}Al MAS NMR analyses were conducted to investigate chemical states of aluminum species in the A-ES and A-AS supports. Although ^{27}Al MAS NMR analyses of NA-ES and NA-AS catalysts were also carried out, significant decrease in NMR signal (poor resolution of peaks) was observed. This is due to the presence of paramagnetic nickel species in the Al_2O_3 supports. It is known that NMR signal is reduced when a paramagnetic metal ion is near the NMR nucleus [32]. Therefore, we obtained NMR spectra of A-ES and A-AS supports to indirectly investigate the effect of preparation method on chemical states of Al atoms in the Al_2O_3 crystalline structures. It is known that chemical states of Al atoms can be classified into three types; octahedral (Al^{VI}), pentahedral (Al^V), and tetrahedral (Al^{IV}) coordinations [33] and [34]. As-synthesized Al_2O_3, which is composed of $Al(OH)_3$, mainly has octahedrally coordinated Al atoms. During the thermal treatment, however, these octahedrally coordinated Al atoms in $Al(OH)_3$ migrate toward tetrahedral or pentahedral coordination sites because of dehydration and dehydroxylation of $Al(OH)_3$ [35]. As shown in Fig. 4, both A-ES and A-AS supports exhibited three peaks at around 8, 37, and 67 ppm, corresponding to octahedrally coordinated Al, pentahedrally coordinated Al, and tetrahedrally coordinated Al, respectively. These NMR spectra were deconvoluted to three bands, and percentages of peak area were calculated as presented Table 2. It is noted that the peak area percentage for pentahedrally coordinated Al in the A-ES support was much larger than that in the A-AS support, representing that a more defective structure was formed in the A-ES support. Because epoxide served as an acid scavenger in the epoxide-driven sol-gel method, gelation process in the epoxide-driven sol-gel method was faster than that in the alkoxide-based sol-gel method. As a result, unsaturated Al atoms were more easily formed in the epoxide-driven sol-gel method than in the alkoxide-based sol-gel method. In other words, disorder in coordination state of Al atoms easily occurred in the epoxide-driven sol-gel method. According to previous study [36], pentahedrally coordinated Al atoms act as Lewis acid center on the surface of -Al_2O_3. When Ni^{2+} ions incorporate into the lattice of Al_2O_3, Lewis acid sites serve as anchoring sites for nickel species. From this, it is believed that large amount of pentahedrally coordinated Al atom as electron-deficient sites was responsible for fine dispersion of nickel species on the surface of NA-ES catalyst. It was also found that chemical shifts of Al atoms in the A-ES support were smaller than those in the A-AS support. This might be due to the less deshielded environment around Al atom in the A-ES support. It is believed that the formation of less deshielded environment around Al atoms in the A-ES support was caused by decrease in amount of high coordination state of Al atoms as represented in Table 2.

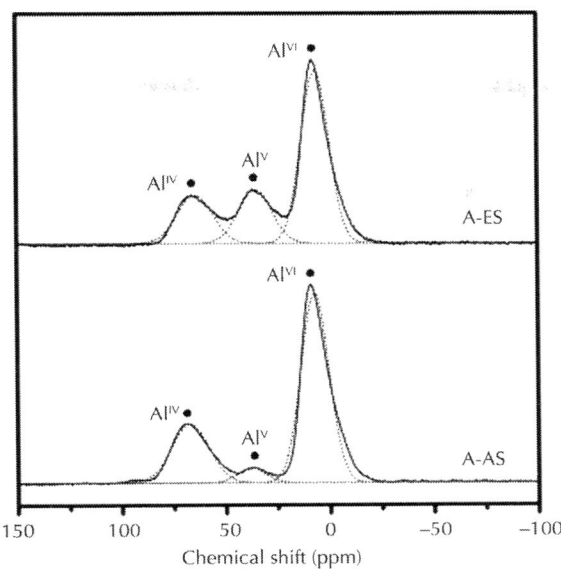

Figure 4: ^{27}Al MAS NMR spectra of A-ES and A-AS supports calcined at 700 °C.

Table 2: Chemical states of aluminum species in the A-ES and A-AS supports calcined at 700 °C

Support	A-ES		A-AS	
	Chemical shift (ppm)	Peak area (%)	Chemical shift (ppm)	Peak area (%)
Al^{VI} (Octahedral)	7.9	57.9	8.8	67.5
Al^{V} (Pentahedral)	36.2	21.5	37.1	4.9
Al^{IV} (Tetrahedral)	66.0	20.6	68.1	27.6

Metal-support Interaction in the Ni-Al$_2$O$_3$ Aerogel Catalysts

TPR measurements were carried out to examine the interaction between nickel species and Al$_2$O$_3$ support in the Ni-Al$_2$O$_3$ aerogel catalysts. Fig. 5 shows the TPR profiles of NA-ES and NA-AS catalysts calcined at 700 °C. It was found that the reduction peak temperature of NA-ES catalyst was lower than that of NA-AS catalyst. In other words, nickel species in the NA-ES catalyst was more reducible than those in the NA-AS catalyst, although nickel loading in the catalysts was nearly identical. It is known that tetrahedral coordination sites or octahedral coordination sites of Al_2O_3 support are occupied by Ni^{2+} ions during the preparation of $Ni-Al_2O_3$ catalysts, resulting in the formation of nickel aluminate phase [37]. Ni^{2+} ions in the octahedral coordination sites of Al_2O_3 support (Ni_{oct}) are more reducible than those in the tetrahedral coordination sites of Al_2O_3 support (Ni_{tet}). From these results, it is inferred that NA-ES catalyst exhibiting a lower reduction temperature mainly consists of more reducible Ni_{oct} compared to NA-AS catalyst. This is because smaller amount of octahedrally coordinated Al atom was formed by the epoxide-driven sol-gel method than by the alkoxide-based sol-gel method (Table 2). Therefore, Ni^{2+} ions can easily occupy cationic vacancies in the octahedral coordination sites of Al_2O_3 support in the NA-ES catalysts.

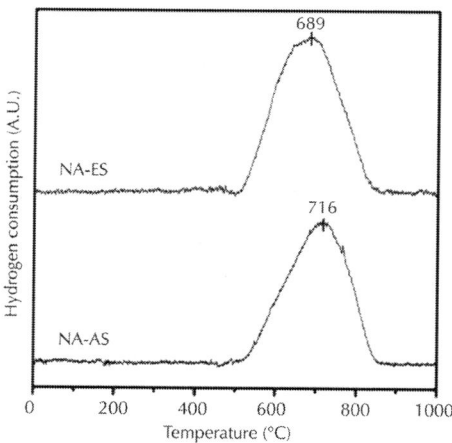

Figure 5: TPR profiles of NA-ES and NA-AS catalysts calcined at 700 °C.

Crystalline Structure of Reduced Ni-Al$_2$O$_3$ Catalysts

XRD patterns of NA-ES and NA-AS catalysts reduced at 700 °C are presented in Fig. 6. It is noticeable that the diffraction peak of -Al$_2$O$_3$ (440) in the calcined catalysts shifted to the higher diffraction angle (Fig. 3 and Fig. 6), i.e., toward original diffraction angle of -Al$_2$O$_3$ (440) phase (dashed line in Fig. 6). Furthermore, the reduced catalysts showed diffraction peaks corresponding to metallic nickel (solid lines inFig. 6). This implies that nickel aluminate phase in both calcined NA-ES and NA-AS catalysts were successfully reduced into metallic nickel during the reduction process employed in this work.

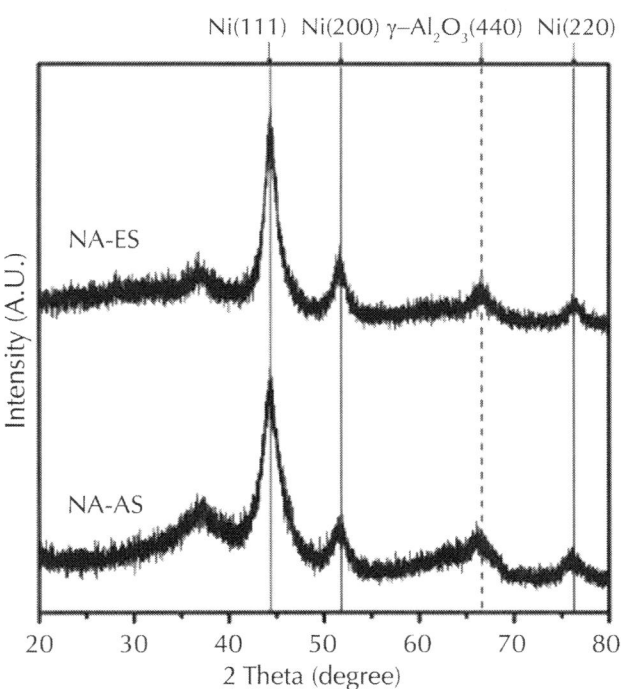

Figure 6: XRD patterns of NA-ES and NA-AS catalysts reduced at 700 °C.

The diffraction angle of -Al_2O_3 (440) in the reduced NA-ES catalyst was almost identical to that of -Al_2O_3, whereas the diffraction angle of -Al_2O_3 (440) appeared at slightly small angle region in the reduced NA-AS catalyst. These observations were quantified as summarized in Table 3, in which lattice parameters of the calcined and reduced catalysts were listed. Although lattice parameter of the calcined NA-ES catalyst was larger than that of the calcined NA-AS catalyst, lattice parameter of the reduced NA-ES catalyst was smaller than that of the reduced NA-AS catalyst. Therefore, it can be inferred that nickel aluminate phase of the calcined NA-ES catalyst was more favorable to be reduced than that of the calcined NA-AS catalyst, as evidenced in the TPR results (Fig. 5). In other words, more reducible nickel aluminate phase was formed through the epoxide-driven sol-gel method.

Table 3: Lattice parameters of NA-ES and NA-AS catalysts

Catalyst	Lattice parameter after calcination (nm)[a]	Lattice parameter after reduction (nm)[b]
NA-ES	0.814	0.793
NA-AS	0.808	0.800

[a]Calculated from Al_2O_3 (440) diffraction peak in Fig. 3.

[b]Calculated from Al_2O_3 (440) diffraction peak in Fig. 6.

Nickel Dispersion of Reduced Ni-Al_2O_3 Catalysts

Hydrogen chemisorption measurements were conducted in order to determine the nickel surface area, nickel dispersion, and average nickel diameter in the reduced NA-ES and NA-AS catalysts. Hydrogen chemisorption results for reduced NA-ES and NA-AS catalysts are listed in Table 4. The amount of hydrogen uptake and nickel surface area of NA-ES catalysts were larger than those of NA-AS catalyst. Consequently, active nickel species were more finely dispersed in the NA-ES catalyst than in the NA-AS catalyst, resulting in smaller average nickel diameter in the NA-ES catalyst. This implies that NA-ES catalyst exhibited stronger resistance toward aggregation of nickel species during the reduction process than NA-AS catalyst.

Table 4: Hydrogen chemisorption results for reduced NA-ES and NA-AS catalysts

Catalyst	NA-ES	NA-AS
Amount of hydrogen uptake (μmol/g-catalyst)[a]	174.4	154.6
Nickel surface area (m^2/g-catalyst)[a]	13.6	12.0
Nickel dispersion (%)[a]	7.2	6.3
Average nickel diameter (nm)[a]	14.1	15.9

[a]Calculated by assuming $H/Ni_{atom} = 1$.

Hydrogen Production by Steam Reforming of LNG over Ni-Al$_2$O$_3$ Aerogel Catalysts

Fig. 7 shows the LNG conversions and hydrogen yields with time on stream in the steam reforming of LNG over NA-ES and NA-AS catalysts at 600 °C. Both NA-ES and NA-AS catalysts showed a stable catalytic performance without any significant catalyst deactivation during the steam reforming reaction. It is believed that internal mass transfer of reactants and products is promoted by well-developed mesopores of the catalysts in the steam reforming reaction. Fine dispersion of active metallic nickel species on the catalysts was also responsible for stable catalytic performance by suppressing carbon deposition and nickel sintering. CHNS analysis revealed that the amounts of carbon deposited on the surface of NA-ES and NA-AS catalysts after a 1000 min-reaction were very small (1.6 and 2.3 wt%, respectively).

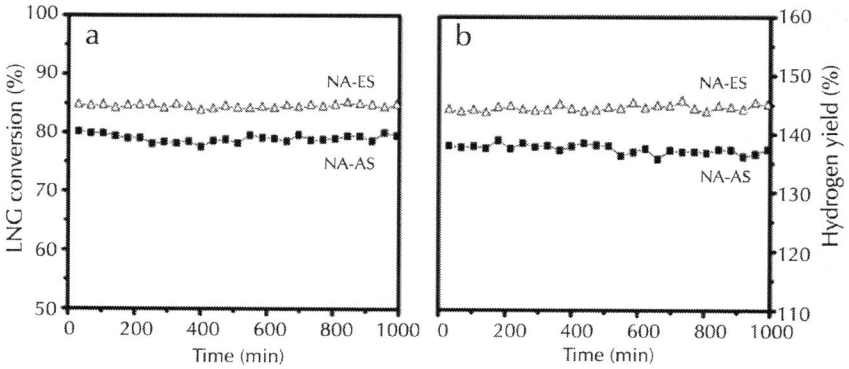

Figure 7: (a) LNG conversions and (b) hydrogen yields with time on stream in the steam reforming of LNG over NA-ES and NA-AS catalysts at 600 °C. All the catalysts were reduced at 700 °C for 3 h prior to the reaction.

NA-ES catalyst exhibited a higher catalytic activity in the steam reforming of LNG than NA-AS catalyst in terms of LNG conversion and hydrogen yield. This result was due to fine dispersion of active nickel species in the NA-ES catalyst. Average nickel diameter served as an important factor determining the catalytic performance in the steam reforming of LNG. According to previous kinetic study on the steam reforming of methane [38], C–H bond cleavage of methane on the surface active nickel site is known as a rate-determining step for overall reaction. It has also been reported that methane is more preferentially adsorbed on active nickel sites where the energy level for dehydrogenated methane can be lowered [39]. This indicates that small nickel particle is favorable for high catalytic activity in the steam reforming reaction, because stronger interaction between active nickel site and reactant is possible on the smaller nickel particle, causing a stabilization effect on the adsorbed intermediate.

Compositions of outlet gas obtained after a 1000 min-reaction over NA-ES and NA-AS catalysts are presented in Fig. 8. Composition of carbon monoxide over NA-ES catalyst was larger than that over NA-AS catalyst, while composition of carbon dioxide over NA-ES catalyst was smaller than that over NA-AS catalyst. This result indicates water-gas shift reaction (equation (4)) was suppressed over NA-ES catalyst during the steam reforming reaction. This is because smaller nickel particles are easier to contact surface of Al_2O_3, resulting a formation of weak bond between carbon monoxide and nickel particles [40]. From this result, it can be inferred that steam reforming of methane (equation (5)) occurred more actively than water-gas shift reaction over the NA-ES catalyst, resulting higher hydrogen yield over the NA-ES catalyst.

$$CO + H_2O \leftrightarrow CO_2 + H_2 \tag{4}$$

$$CH_4 + H_2O \leftrightarrow CO + 3H_2 \tag{5}$$

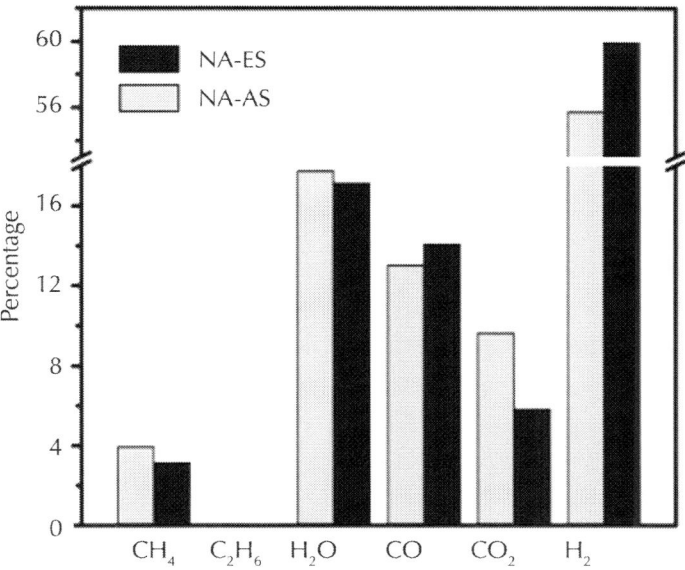

Figure 8: Compositions of outlet gas obtained after a 1000 min-reaction over NA-ES and NA-AS catalysts.

It is concluded that NA-ES catalyst with small average nickel diameter served as an efficient catalyst for hydrogen production by steam reforming of LNG. Small nickel particles of NA-ES catalyst were favorable for strong interaction between reactant and catalyst surface. High reducibility of nickel species in the NA-ES catalyst was also partly responsible for excellent catalytic performance in the steam reforming reaction.

CONCLUSIONS

A mesoporous Ni-Al$_2$O$_3$ aerogel catalyst was prepared by a single-step epoxide-driven sol-gel method and a subsequent supercritical CO$_2$ drying method (NA-ES catalyst). To examine the difference in physicochemical properties and catalytic activities of Ni-Al$_2$O$_3$ catalysts in the steam reforming of LNG, a mesoporous Ni-Al$_2$O$_3$ aerogel catalyst was also prepared by a single-step alkoxide-based sol-gel method and a subsequent supercritical CO$_2$ drying method (NA-AS catalyst). It was found that pore structure of the catalysts was influenced by the preparation method. Epoxide-driven sol-gel method was favorable for fine dispersion of nickel species, because large amount of pentahedral Al atoms served as Lewis acid sites for anchoring nickel species. In addition, NA-ES catalyst retained high reducibility and finely dispersed nickel particles compared to NA-AS catalyst. Both NA-ES and NA-AS catalysts showed a stable catalytic performance during the steam reforming reaction without any significant deactivation. However, NA-ES catalyst with small average nickel diameter exhibited higher LNG conversion and hydrogen yield in the steam reforming reaction than NA-AS catalyst. Furthermore, water-gas shift reaction was suppressed over NA-ES catalyst. In conclusion, NA-ES catalyst prepared by an epoxide-driven sol-gel method served as an efficient catalyst in the steam reforming of LNG.

ACKNOWLEDGMENTS

The authors wish to acknowledge support from Korean Ministry of Education, Science and Technology (Global Frontier Project).

REFERENCES

1.   Winter C-J. Hydrogen energy - Abundant, efficient, clean: a debate over the energy system-of-change. Int J Hydrogen Energy 2009;34:S1-52.

2.   Schrope M. Which way to energy utopia? Nature 2001;414: 682-4.

3.   Dunn S. Hydrogen futures: toward a sustainable energy system. Int J Hydrogen Energy 2002;27:235-64.

4.   Cheekatamarla PK, Finnerty CM. Reforming cata ysts for hydrogen generation in fuel cell applications. J Power Sources 2006;160:490-9.

5.   Armor JN. The multiple roles for catalysis in the production of H2. Appl Catal A Gen 1999;176:159-76.

6.   Navarro RM, Pen~ a MA, Fierro JLG. Hydrogen production reactions from carbon feedstocks: fossil fuels and biomass. Chem Rev 2007;107:3952-91.

7.   Seo JG, Youn MH, Cho KM, Park S, Lee SH, Lee J, et al. Effect of Al2O3-ZrO2 xerogel support on the hydrogen production by steam reforming of LNG over Ni/Al2O3-ZrO2 catalyst. Korean J Chem Eng 2008;25:41-5.

8.   Haryanto A, Fernando S, Murali N, Adhikari S. Current status of hydrogen production techniques by steam reforming of ethanol: a review. Energy Fuels 2005;19:2098-106.

9.   Meng F, Chen G, Wang Y, Liu Y. Metallic Ni monolith-Ni/MgAl2O4 dual bed catalysts for the autotherma partial oxidation of methane to synthesis gas. Int J Hydrogen Energy 2010;35:8182-90.

10. Enger BC, Lødeng R, Holmen A. A review of catalytic partial oxidation of methane to synthesis gas with emphasis on reaction mechanisms over transition metal catalysts. Appl Catal A Gen 2008;346:1-27.

11. Gao J, Guo J, Liang D, Hou Z, Fei J, Zheng X. Production of syngas via autothermal reforming of methane in a fluidizedbed

reactor over the combined CeO2-ZrO2/SiO2 supported Ni catalysts. Int J Hydrogen Energy 2008;33:5493-500.

12. Youn MH, Seo JG, Cho KM, Jung JC, Kim H, La KW, et al. Effect of support on hydrogen production by auto-thermal reforming of ethanol over supported nickel catalysts. Korean J Chem Eng 2008;25:236-8.

13. Garcı́a-Die´guez M, Pieta IS, Herrera MC, Larrubia MA, Alemany LJ. Nanostructured Pt- and Ni-based catalysts for CO2-reforming of methane. J Catal 2010;270:136-45.

14. Rostrup-Nielsen JR, Sehested J, Nørskov JK. Hydrogen and synthesis gas by steam- and CO2 reforming. Adv Catal 2002; 47:65-139.

15. Rostrup-Nielsen JR. New aspects of syngas production and use. Catal Today 2000;63:159-64.

16. Bengaard HS, Nørskov JK, Sehested J, Clausen BS, Nielsen LP, Molenbroek AM, et al. Steam reforming and graphite formation on Ni catalysts. J Catal 2002;209:365-84.

17. Salhi N, Petit C, Kiennemann A. Steam reforming of methane on nickel aluminate defined structures with high Al/Ni ratio. Stud Surf Sci Catal 2008;174:1335-8.

18. Sehested J. Four challenges for nickel steam-reforming catalysts. Catal Today 2006;111:103-10.

19. Maluf SS, Assaf EM. Ni catalysts with Mo promoter for methane steam reforming. Fuel 2009;88:1547-53.

20. Seo JG, Youn MH, Park S, Lee J, Lee SH, Lee H, et al. Hydrogen production by steam reforming of LNG over Ni/Al2O3-ZrO2 catalysts: effect of ZrO2 and preparation method of Al2O3-ZrO2. Korean J Chem Eng 2008;25:95-8.

21. Kim H-W, Kang K-M, Kwak H-Y, Kim J- H. Preparation of supported Ni catalysts on various metal oxides with core/shell structures and their tests for the steam reforming of methane. Chem Eng J 2011;168:775-83.

22. Seo JG, Youn MH, Jung JC, Song IK. Hydrogen production by steam reforming of liquefied natural gas (LNG) over

mesoporous nickel-alumina aerogel catalyst. Int J Hydrogen Energy 2010;35:6738-46.

23. Bang Y, Seo JG, Song IK. Hydrogen production by steam reforming of liquefied natural gas (LNG) over mesoporous NiLa-Al2O3 aerogel catalysts: effect of La content. Int J Hydrogen Energy 2011;36:8307-15.

24. Yoldas BE. A transparent porous alumina. Amer Ceram Soc Bull 1975;54:286-8.

25. Gash AE, Tillotson TM, Satcher JH, Poco JF, Hrubesh LW, Simpson RL. Use of epoxides in the sol-gel synthesis of porous iron(III) oxide monoliths from Fe(III) salts. Chem Mater 2001;13:999-1007.

26. Baumann TF, Gash AE, Chinn SC, Sawvel AM, Maxwell RS, Satcher JH. Synthesis of high-surface-area alumina aerogels without the use of alkoxide precursors. Chem Mater 2005;17: 395-401.

27. Chen L, Zhu J, Liu Y-M, Cao Y, Li H-X, He H-Y, et al. Photocatalytic activity of epoxide sol-gel derived titania transformed into nanocrystalline aerogel powders by supercritical drying. J Mol Catal A Chem 2006;255:260-8.

28. Gonza´lez-Pena V, Marquez-Alvarez C, Di´az I, Grande M, Blasco T, Pe´rez-Pariente J. Sol-gel synthesis of mesostructured aluminas from chemically modified aluminum sec-butoxide using non-ionic surfactant templating. Micropor Mesopor Mater 2005;80:173-82.

29. Sing KSW, Everett DH, Haul RAW, Moscou L, Pierotti RA, Rouquerol J, et al. Reporting physisorption data for gas/solid systems. Pure Appl Chem 1985;57:603-19.

30. Kim P, Kim Y, Kim H, Song IK, Yi J. Preparation, characterization, and catalytic activity of NiMg catalysts supported on mesoporous alumina for hydrodechlorination of o-dichlorobenzene. J Mol Catal A Chem 2005;231:247-54.

31. Dimotakis ED, Pinnavaia TJ. New route to layered double hydroxides intercalated by organic anions: precursors to

polyoxometalate-pillared derivatives. Inorg Chem 1990;29: 2393-4.

32.   Zhang W, Sun M, Prins R. A high-resolution of MAS NMR study of the structure of fluorinated NiW/g-Al2O3 hydrotreating catalysts. J Phys Chem B 2003;107:10977-82.

33.   Onfroy T, Li W-C, Schu¨th F, Kno¨zinger H. Surface chemistry of carbon-templated mesoporous aluminas. Phys Chem Chem Phys 2009;11:3671-9.

34.   Kim Y, Kim C, Kim P, Yi J. Effect of preparation conditions on the phase transformation of mesoporous alumina. J NonCryst Solids 2005;351:550-6.

35.   Krokidis X, Raybaud P, Gobichon A-E, Rebours B, Enzen R, Toulhoat H. Theoretical study of the dehydration process of boehmite to gamma-alumina. J Phys Chem B 2001;105: 5121-30.

36.   Kwak JH, Hu JZ, Kim DH, Szanyi J, Peden CHF. Pentacoordinated Al3þ ions as preferential nucleation sites for BaO on g-Al2O3: an ultra-high-magnetic field 27Al MAS NMR study. J Catal 2007;251:189-94.

37.   Wu M, Hercules DM. Studies of supported nickel catalysts by X-ray photoelectron and ion scattering spectroscopies. J Phys Chem 1979;83:2003-8.

38.   Trimm DL. The steam reforming of natural gas: problems and some solutions. Stud Surf Sci Catal 1987;36:39-50.

39.   Rostrup-Nielsen JR, Nørskov JK. Step sites in syngas catalysis. Top Catal 2006;40:1-4.

40.   Bartholomew CH, Pannell RB. The stoichiometry of hydrogen and carbon monoxide chemisorption on alumina- and silicasupported nickel. J Catal 1980; 65:390-401.

Hydrogen Production by Steam Reforming of Liquefied Natural Gas (LNG) Over Mesoporous Nickel–alumina Xerogel Catalysts: Effect of Nickel Content

Jeong Gil Seo[a], Min Hye Youn[a], Ho-In Lee[a], Jae Jeong Kim[a], Eunsun Yang[b], Jin Suk Chung[c], Pil Kim[d], and In Kyu Song[a]

[a]School of Chemical and Biological Engineering, Research Center for Energy Conversion and Storage, Seoul National University, Shinlim-dong, Kwanak-ku, Seoul 151-744, Republic of Korea

bDepartment of Chemical Engineering and Applied Chemistry, University of Toronto, 200 College Street, Toronto, ON M5S 3E5, Canada

cSchool of Chemical Engineering and Bioengineering, University of Ulsan, Ulsan 680-749, Republic of Korea

dSchool of Environmental and Chemical Engineering, Chonbuk National University, Jeonju, Chonbuk 561-756, Republic of Korea

ABSTRACT

Mesoporous nickel–alumina xerogel (XNiAl) catalysts with various nickel contents were prepared by a single-step sol–gel method for use in hydrogen production by steam reforming of liquefied natural gas (LNG). The effect of nickel content on the catalytic performance of XNiAl catalysts was investigated. Nickel species were finely dispersed in the XNiAl catalysts through the formation of Ni–O–Al composite structure. The XNiAl catalysts served as efficient catalysts in the hydrogen production by steam reforming of LNG. Both LNG conversion and H_2 composition in dry gas showed volcano-shaped curves with respect to nickel content. Thus, optimal nickel content was required for maximum catalytic performance. The performance of XNiAl catalysts in the steam reforming of LNG increased with increasing reducibility of the catalyst. Among the catalysts examined, the 30NiAl (30 wt% Ni) catalyst with the highest reducibility showed the best catalytic performance. The highest surface area and the largest pore volume of the 30NiAl (30 wt% Ni) catalyst were also partly responsible for its superior catalytic performance.

INTRODUCTION

Hydrogen has attracted much attention as an alternative energy source due to its clean, renewable, and non-polluting nature [1]. Technological advances in hydrogen utilization such as fuel cell make hydrogen more important as a new energy source. However, development of feasible production methods for hydrogen is necessary, because abundant hydrogen is not given to us in nature but should be produced from water or organic compounds [2]. A number of catalytic reforming technologies, such as steam reforming, partial oxidation, and auto-thermal reforming, have been extensively investigated for the large and small scale hydrogen production from various hydrocarbons [3], [4], [5], [6], [7], [8], [9], [10] and [11]. Among the reforming technologies, steam reforming of methane has been recognized as a feasible route to produce hydrogen. Liquefied natural gas (LNG), which is abundant and mainly composed of methane, can serve as an alternate source for hydrogen production by steam reforming reaction. The extensive piping system for LNG in modern cities also makes LNG well suited as a hydrogen source for residential reformers in fuel cell applications.

Nickel-based catalysts have been widely studied as efficient catalysts for various reactions, such as hydrogenation of unsaturated hydrocarbons and reforming of hydrocarbons [12], [13], [14], [15] and [16]. In particular, Ni/Al_2O_3 catalysts have been recognized as promising catalysts for steam reforming reactions due to their low cost and high catalytic activity [17], [18], [19], [20] and [21]. The Ni/Al_2O_3 catalysts, however, require high reaction temperatures and excess amounts of steam to prevent sintering of nickel particles and deposition of carbon species on the catalyst surface in the stream reforming reactions [3], [9] and [22].

The catalytic activity of Ni/Al_2O_3 is closely related to both nickel content and nickel dispersion, but these two factors have opposite effects on the catalytic activity. With increasing nickel content, for example, the catalytic activity of Ni/Al_2O_3 increases due to the increased number of active nickel sites, but the dispersion of nickel particles decreases due to the aggregation of nickel species. In general, nickel content of conventional Ni/Al_2O_3 catalysts used in the steam reforming reactions does not exceed 12 wt% to avoid severe aggregation or sintering of nickel particles during the reactions [23]. Although stable Ni/Al_2O_3 catalysts can be obtained by lowering the nickel content, they may show an inferior catalytic activity due to the insufficient number of active nickel sites. Furthermore, the nickel catalysts that are highly dispersed on Al_2O_3 readily form nickel aluminate phases through the incorporation of Ni^{2+} into the lattice of Al_2O_3 [21], [24] and [25]. The strong metal–support interaction, in turn, inhibits the reduction of nickel aluminate into active metallic nickel.

Many attempts have been made to increase the stability of Ni/ Al_2O_3 catalysts in the steam reforming reactions [17], [20], [21], [26], [27] and [28]. The performance of Ni/Al_2O_3 catalysts in the steam reforming reactions depends not only on the nature and structure of active nickel, but also on the chemical and physical properties of Al_2O_3. It is known that metal oxides prepared by a sol–gel method retain hydroxyl-rich surfaces, and therefore, exhibit unique chemical and physical properties compared to those prepared by a conventional method. In particular, alumina materials prepared by a sol–gel method have high-surface areas and controllable chemical and physical properties. It has been reported that a nickel–alumina xerogel catalyst prepared by a sol–gel method inhibited carbon deposition in the dry reforming of methane, resulting in the enhanced methane conversion and coke resistance [29] and [30]. Therefore, developing a sol–gel derived nickel–alumina catalyst, which retains both high activity and stability in the steam reforming of LNG, would be worthwhile.

In this work, a series of mesoporous nickel–alumina xerogel catalysts with various nickel contents were prepared by a single-step sol–gel method for use in hydrogen production by steam reforming of LNG. The effect of nickel content on the catalytic performance of mesoporous nickel–alumina xerogel catalysts was investigated. It is expected that the mesoporous nickel–alumina xerogel (XNiAl) catalysts prepared by a single-step sol–gel method would show a high and stable catalytic performance in the steam reforming of LNG without significant nickel sintering and carbon deposition.

EXPERIMENTAL

Preparation of Mesoporous Nickel–alumina Xerogel Catalysts

A series of mesoporous nickel–alumina xerogel catalysts with various nickel contents were prepared by a single-step sol–gel method, according to the similar method reported in the literatures [29], [30] and [31]. A known amount of aluminum precursor (aluminum sec-butoxide, Sigma–Aldrich) was dissolved in ethanol at 80 °C with vigorous stirring. Small amounts of distilled water and nitric acid, which had been diluted with ethanol, were slowly added into the solution of aluminum precursor for the partial hydrolysis of the aluminum precursor. After maintaining the resulting solution at 80 °C for a few minutes, a clear sol was obtained. The sol was cooled to 60 °C, and then a known amount of nickel precursor (nickel acetate tetrahydrate, Sigma–Aldrich) was slowly added into the sol to obtain a nickel–alumina sol. After cooling the nickel–alumina sol to room temperature, a monolithic gel was obtained by adding an appropriate amount of water diluted with ethanol into the sol. The gel was aged for 7 days, and then dried overnight at 120 °C. The resulting powder was finally calcined at 700 °C for 5 h to yield the mesoporous nickel–alumina xerogel catalyst. The prepared nickel–alumina xerogel catalysts were denoted as XNiAl (X = 15, 20, 25, 30, 35, and 40), where X represents the nickel content (wt %) in the catalyst. For example, 30NiAl denotes a 30-wt% nickel–alumina xerogel catalyst.

For the purpose of comparison, a nickel catalyst supported on commercial Al_2O_3 (Degussa) was prepared by an impregnation method. The nickel loading was fixed at 20 wt%. The supported nickel catalyst was denoted as 20Ni/Al_2O_3-impregnation.

Characterization

Nitrogen adsorption–desorption isotherms of the catalysts were obtained with an ASAP-2010 (Micromeritics) instrument. Average pore diameters of the catalysts were determined by the Barret–Joyner–Hallender (BJH) method applied to the desorption branch of the nitrogen isotherm. Nickel dispersion on the catalysts was examined by TEM analyses (Jeol, JEM-2000EXII). Crystalline phases of the catalysts were investigated by XRD (MAC Science, M18XHF-SRA) measurements using Cu K radiation (= 1.54056 Å) operated at 50 kV and 100 mA. In order to examine the reducibility of the catalysts, temperature-programmed reduction (TPR) measurements were carried out in a conventional flow system with a moisture trap connected to a thermal conductivity detector (TCD) at temperatures ranging from room temperature to 1000 °C with a ramping rate of 5 °C/min. For the TPR measurements, a mixed stream of H_2 (2 ml/min) and N_2 (20 ml/min) was used for 0.2 g of catalyst sample.

Steam Reforming of LNG

Steam reforming of LNG was carried out in a continuous flow fixed-bed reactor at atmospheric pressure. Each calcined catalyst (100 mg) was charged into a tubular quartz reactor, and then reduced with a mixed stream of H_2 (3 ml/min) and N_2 (30 ml/min) at 700 °C for 3 h. Water was sufficiently vaporized and continuously fed into the reactor together with LNG (92.0 vol. % CH_4 and 8.0 vol. % C_2H_6) and N_2 carrier (30 ml/min). The steam/carbon ratio in the feed stream was fixed at 2.0, and the total feed rate with respect to the catalyst was maintained at 27,000 ml h^{-1}/g. The catalytic reaction was carried out at 600 °C. Reaction products were periodically sampled and analysed using an on-line gas chromatograph (Younglin, ACME 6000) equipped with a thermal conductivity detector. LNG conversion and H_2 composition in dry gas were calculated on the basis of carbon balance as follows

$$LNG \text{ conversion } (\%) = \left(1 - \frac{F_{CH_4,out} + F_{C_2H_6,out}}{F_{CH_4,in} + F_{C_2H_6,in}}\right) \times 100 \tag{1}$$

H_2 composition in dry gas $(\%)$

$$= \frac{F_{H_2,out}}{\begin{array}{c} F_{H_2,out} + F_{CH_4,out} + F_{C_2H_6,out} \\ + F_{CO,out} + F_{CO_2,out} \end{array}} \times 100 \tag{2}$$

RESULTS AND DISCUSSION

Chemical and Physical Property of XNiAl Xatalysts

Physical properties of XNiAl (X = 15, 20, 25, 30, 35, and 40) catalysts were examined by nitrogen adsorption–desorption isotherm measurements. Fig. 1 shows the nitrogen adsorption–desorption isotherms and pore size distributions of selected XNiAl (X = 15, 30, and 40) catalysts. The XNiAl (X = 15, 30, and 40) catalysts showed IV-type isotherms with H_2-type hysteresis loops, indicating the existence of well-developed framework mesopores. Furthermore, the XNiAl (X = 15, 30, and 40) catalysts showed narrow pore size distributions centered at around 2–5 nm. The XNiAl (X = 20, 25, and 35) catalysts also showed isotherms, hysteresis loops, and pore size distributions similar to those observed for XNiAl (X = 15, 30, and 40) catalysts. These results indicate that mesoporous nickel–alumina xerogel catalysts were successfully prepared in this work.

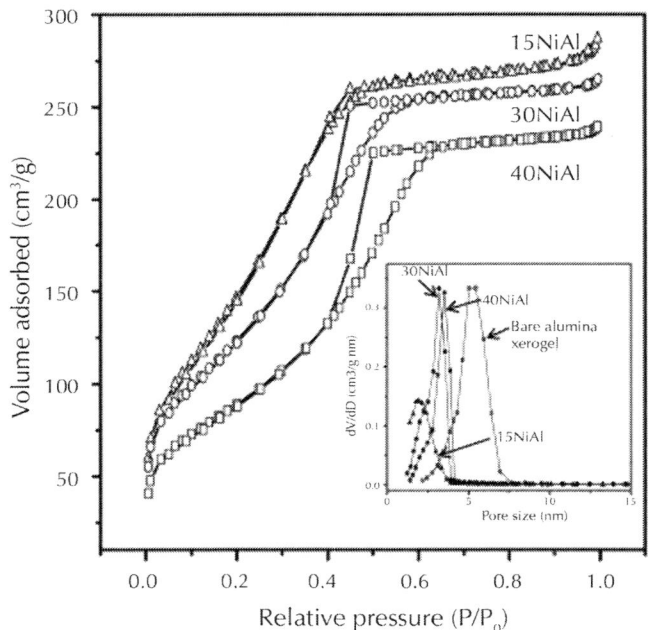

Figure 1: Nitrogen adsorption–desorption isotherms and pore size distributions of selected XNiAl (X = 15, 30, and 40) catalysts. All the catalysts were calcined at 700 °C prior to the measurements.

Detailed chemical and physical properties of XNiAl (X = 15, 20, 25, 30, 35, and 40) catalysts are summarized in Table 1. For comparison, the physical properties of bare alumina xerogel [31] are also listed in Table 1. It was revealed that the nickel contents of XNiAl (X = 15, 20, 25, 30, 35, and 40) catalysts were almost identical to the target nickel loadings. The incorporation of nickel into mesoporous alumina did not greatly affect the surface area, but decreased the pore volume and average pore diameter of XNiAl (X = 15, 20, 25, 30, 35, and 40) catalysts, when compared to the bare alumina xerogel. It was found that the surface area and the pore volume of the catalysts showed volcano-shaped curves with respect to nickel content, although they did not show the same trend. Among the prepared catalysts, the 30NiAl catalyst exhibited the highest surface area and the largest pore volume.

Table 1: Chemical and physical properties of XNiAl (X = 15, 20, 25, 30, 35, and 40) catalysts calcined at 700 °C for 5 h

Catalyst	Ni/Al atomic ratio[a]	Actual Ni loading (wt%)[a]	Surface area (m^2 g^{-1})[b]	Pore volume (cm^3 g^{-1})[c]	Average pore diameter (nm)[d]
Bare alumina xerogel	0.0	–	365	0.64	4.7
15NiAl	0.39	17.1 (15)	340	0.20	2.1
20NiAl	0.52	21.7 (20)	360	0.30	2.1
25NiAl	0.64	25.2 (25)	412	0.32	2.4
30NiAl	0.81	30.0 (30)	458	0.42	2.6
35NiAl	0.97	33.9 (35)	333	0.37	3.1
40NiAl	1.14	37.7 (40)	328	0.36	3.0

[a] Determined by ICP-AES analysis (values in parentheses represent the target nickel loading).

[b] Calculated by the BET equation.

[c] BJH desorption pore volume.

[d] BJH desorption average pore diameter.

Nickel Dispersion

Fig. 2 shows the XRD patterns of XNiAl (X = 15, 20, 25, 30, 35, and 40) catalysts calcined at 700 °C for 5 h. For comparison, the XRD pattern of bare alumina xerogel [31] is also presented in Fig. 2. It is noteworthy that no diffraction peaks corresponding to nickel oxide were observed even in the 40NiAl catalyst. This result indicates that nickel species were finely dispersed in the XNiAl (X = 15, 20, 25, 30, 35, and 40) catalysts, resulting in the formation of small nickel particles that were under the detection limit of XRD measurement [32] and [33]. Instead, the diffraction peaks (solid lines) indicative of spinel nickel aluminate phase were observed in all the XNiAl (X = 15, 20, 25, 30, 35, and 40) catalysts. It was difficult to distinguish alumina phase from nickel aluminate phase due to the overlap of XRD peaks. The formation of nickel aluminate phase caused the lattice expansion of alumina because the ionic radius of Ni is larger than that of Al. In this work, it was observed that the (4 4 0) diffraction peak of alumina shifted to lower angle with increasing nickel content. The above result implies that a nickel aluminate phase was formed in the XNiAl (X = 15, 20, 25, 30, 35, and 40) catalysts due to the homogeneous mixing and interaction between alumina sol and nickel precursor during the catalyst preparation. It is likely that the anionic acetate groups in the nickel precursor (nickel acetate tetrahydrate) acted as chelating agents retarding hydrolysis and condensation of the alumina sol, which resulted in the formation of Ni–O–Al composite structure [30].

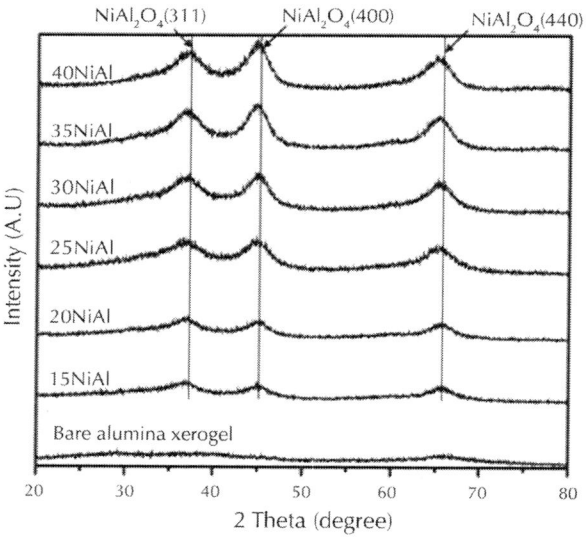

Figure 2: XRD patterns of XNiAl (X = 15, 20, 25, 30, 35, and 40) catalysts calcined at 700 °C for 5 h.

Fine dispersion of nickel species in the prepared catalysts was further confirmed by TEM analyses. Fig. 3shows the TEM images of 30NiAl and 40NiAl catalysts calcined at 700 °C for 5 h. No visible evidence representing nickel agglomerates was found in both catalysts. This result indicates that nickel species were finely dispersed in the nickel–alumina xerogel catalysts, as confirmed by XRD measurements (Fig. 2). TEM images also showed that both

30NiAl and 40NiAl catalysts retained well-developed mesopores, as demonstrated in Fig. 1 and Table 1.

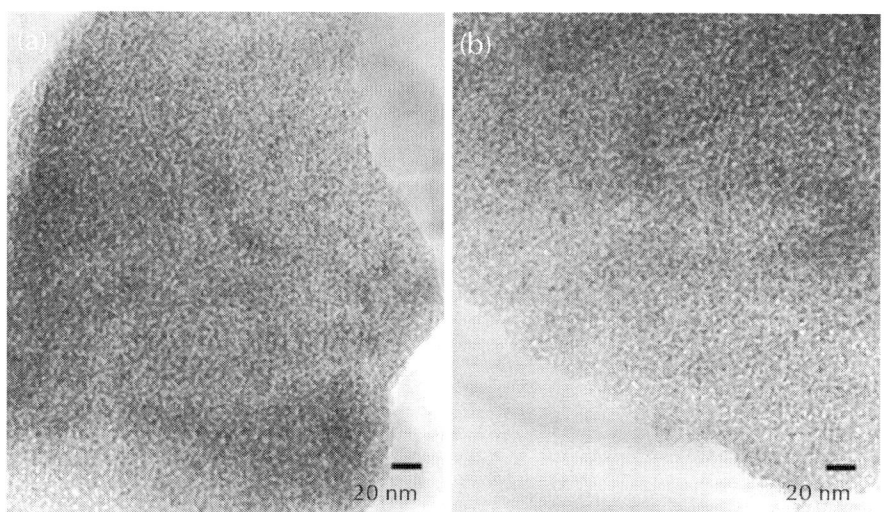

Figure 3: TEM images of (a) 30NiAl and (b) 40NiAl catalysts calcined at 700 °C for 5 h.

Reducibility

TPR measurements were carried out to investigate the reducibility of XNiAl (X = 15, 20, 25, 30, 35, and 40) catalysts, and to examine the interaction between nickel species and alumina. Fig. 4 shows the TPR profiles of XNiAl (X = 15, 20, 25, 30, 35, and 40) catalysts. All XNiAl (X = 15, 20, 25, 30, 35, and 40) catalysts showed a broad reduction band at around 800 °C. This means that the stable nickel aluminate phase was formed in the XNiAl (X = 15, 20, 25, 30, 35, and 40) catalysts, in good agreement with the XRD results (Fig. 2). However, the reduction peak shifted to a lower temperature with increasing nickel content in the XNiAl (X = 15, 20, 25, and 30) catalysts. In other words, the interaction between nickel species and alumina decreased with increasing nickel content in the XNiAl (X = 15, 20, 25, and 30) catalysts. It is believed that the surface nickel aluminate phase, which is easier to be reduced than the bulk nickel aluminate phase, was preferentially formed with increasing nickel content in the XNiAl (X = 15, 20, 25, and 30) catalysts[34] and [35]. On the other hand, both 35NiAl and 40NiAl catalysts exhibited a higher reduction peak temperature than the 30NiAl catalyst. It was reported that the reducibility of nickel catalyst supported on alumina decreased with decreasing nickel loading and with increasing calcination temperature [36]. In the 35NiAl and 40NiAl catalysts, however, the reducibility decreased with increasing nickel content. It is believed that the bulk nickel aluminate phase was dominantly formed in the 35NiAl and 40NiAl catalysts unlike in the XNiAl (X = 15, 20, 25, and 30) catalysts. The nickel species mainly exist on the surface of alumina in the nickel catalyst impregnated on alumina. In the XNiAl (X = 15, 20, 25, 30, 35, and 40) catalysts, however, the nickel species exist both on the surface and in the bulk of alumina due to homogeneous mixing of nickel precursor and alumina sol, resulting in the formation of both surface nickel aluminate phase and bulk nickel aluminate phase. It is believed that the majority of nickel species was located in the bulk of alumina when the excess amount of Ni was added into alumina sol as the case of 35NiAl and 40NiAl catalysts. The TPR results presented in Fig. 4 clearly demonstrate that the reduction peak temperature of XNiAl (X = 15, 20, 25, 30, 35, and 40) catalysts showed a volcano-shaped curve with respect to nickel content. This means that optimal nickel content was required for the effective formation of surface nickel aluminate phase in the XNiAl (X = 15, 20, 25, 30, 35, and 40) catalysts. Among the catalysts examined, the 30NiAl catalyst showed the highest reducibility (the lowest reduction peak temperature).

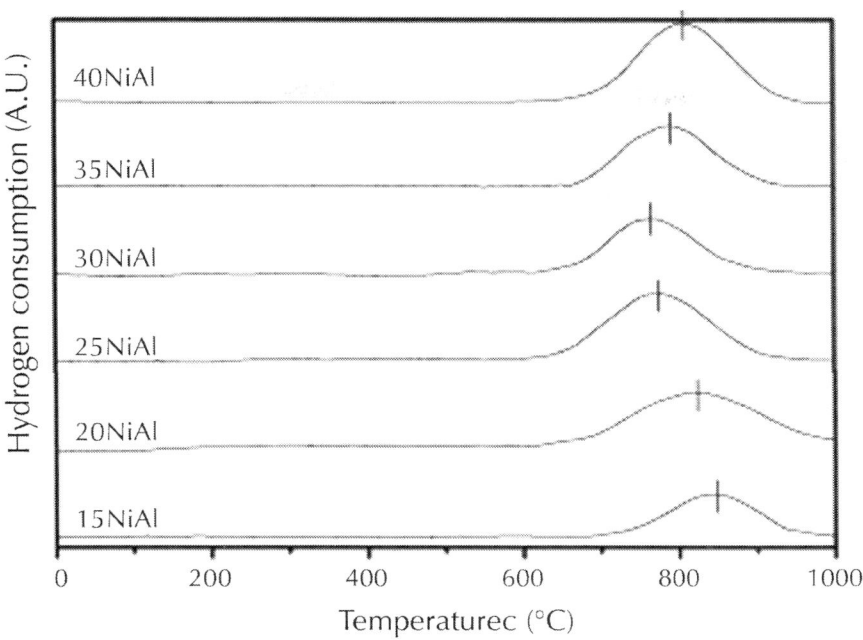

Figure 4: TPR profiles of XNiAl (X = 15, 20, 25, 30, 35, and 40) catalysts calcined at 700 °C for 5 h.

Steam Reforming of LNG

The catalytic performance of XNiAl (X = 15, 20, 25, 30, 35, and 40) and Ni/Al$_2$O$_3$-impregnation catalysts in the steam reforming of LNG at 600 °C is summarized in Table 2. It is well known that the steam reforming of methane is an equilibrium-controlled reaction. The equilibrium methane conversion and hydrogen composition in dry gas at 600 °C were 99.89 and 76.08%, respectively. The performance of XNiAl (X = 15, 20, 25, 30, 35, and 40) and 20Ni/Al$_2$O$_3$-impregnation catalysts was lower than the equilibrium value. However, the H$_2$/CO ratio was higher than the theoretical value.

Table 2: Catalytic performance of XNiAl (X = 15, 20, 25, 30, 35, and 40) and Ni/Al$_2$O$_3$-impregnation catalysts in the steam reforming of LNG at 600 °C

Catalyst	LNG conversion (%)[a]	H$_2$ composition in dry gas (%)[a]	CO composition in dry gas (%)[a]	CO$_2$ composition in dry gas (%)[a]	H$_2$/CO ratio[a]	Carbon deposition (wt%)[b]
Equilibrium at 600 °C	99.89	76.08	16.98	6.28	4.48	–
20Ni/Al$_2$O$_3$-impregnation	15.42	43.98	8.32	9.61	5.29	12.0
15NiAl	66.59	65.23	11.67	15.00	5.58	0.1
20NiAl	72.54	66.07	12.14	14.46	5.44	0.4
25NiAl	79.70	67.45	12.89	14.41	5.22	0.5
30NiAl	80.68	67.70	12.09	15.06	5.60	0.3
35NiAl	78.38	67.24	12.44	14.49	5.40	0.5
40NiAl	76.28	66.52	12.75	14.49	5.22	0.2

[a] Obtained after a 400-min reaction at 600 °C.

[b] Determined by CHNS elemental analysis after a 1000-min reaction at 600 °C.

Fig. 5 shows the LNG conversions with time on stream over 20NiAl, 30NiAl, and 20Ni/Al$_2$O$_3$-impregnation catalysts in the steam reforming of LNG at 600 °C. It is noticeable that the 20Ni/Al$_2$O$_3$-impregnation catalyst experienced a severe catalyst deactivation due to the significant carbon deposition on the catalyst surface. CHNS elemental analyses revealed that the 20Ni/Al$_2$O$_3$-impregnation catalyst contained 12 wt% carbon species after a 1000-min reaction. However, both 20NiAl and 30NiAl catalysts showed a stable catalytic performance during the reaction, extending over 1000 min. It was also observed that the XNiAl (X = 15, 25, 35, and 40) catalysts exhibited a stable catalytic performance during the reaction (although their catalytic performance was not shown in Fig. 5). The amount of carbon species deposited on the XNiAl (X =

15, 20, 25, 30, 35, and 40) catalysts after a 1000-min reaction was less than 0.5 wt%.

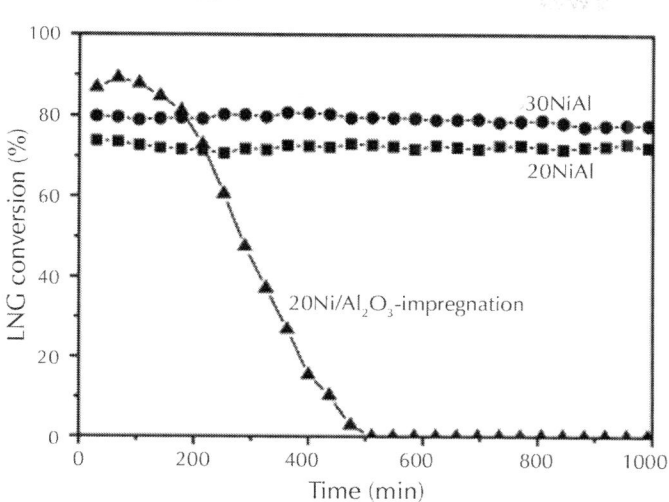

Figure 5: LNG conversions with time on stream over 20NiAl, 30NiAl, and 20Ni/Al$_2$O$_3$-impregnation catalysts in the steam re-forming of LNG at 600 °C. All catalysts were reduced at 700 °C prior to the reaction.

The superior catalytic performance of the XNiAl (X = 15, 20, 25, 30, 35, and 40) catalysts compared to the 20Ni/Al$_2$O$_3$-impregnation catalyst can be explained by the chemical and physical properties of the XNiAl (X = 15, 20, 25, 30, 35, and 40) catalysts. It is believed that the highly dispersed nickel species and the well-developed mesopores in the XNiAl (X = 15, 20, 25, 30, 35, and 40) catalysts greatly enhanced coke resistance by preventing polymerization of the adsorbed surface hydrocarbons during the steam reforming reaction. Furthermore, the strong interaction between nickel species and alumina in the XNiAl (X = 15, 20, 25, 30, 35, and 40) catalysts effectively suppressed sintering of nickel particles through the formation of a stable nickel aluminate phase.

Fig. 6 shows the LNG conversions and H$_2$ compositions in dry gas over XNiAl (X = 15, 20, 25, 30, 35, and 40) catalysts in the steam reforming of LNG, plotted as a function of nickel content. Both LNG conversion and H_2composition in dry gas showed volcano-shaped curves with respect to nickel content, and decreased in the order of 30NiAl > 25NiAl > 35NiAl > 40NiAl > 20NiAl > 15NiAl. This result indicates that optimal nickel content was required for maximum catalytic performance of mesoporous nickel–alumina xerogel catalysts. It is also believed that the highest surface area and the largest pore volume of 30NiAl catalyst were partly responsible for its superior catalytic activity (Table 1). The high surface area and large pore volume of the 30NiAl catalyst improved adsorption of both LNG and steam onto the catalyst surface. Harmonious adsorption of hydrocarbons (LNG) and steam onto the surface of the 30NiAl catalyst, in turn, enhanced the gasification reaction between two components. The surface nickel aluminate phase in the 30NiAl catalyst also played an important role in increasing the surface area of active nickel, resulting in the enhanced catalytic performance of the 30NiAl catalyst.

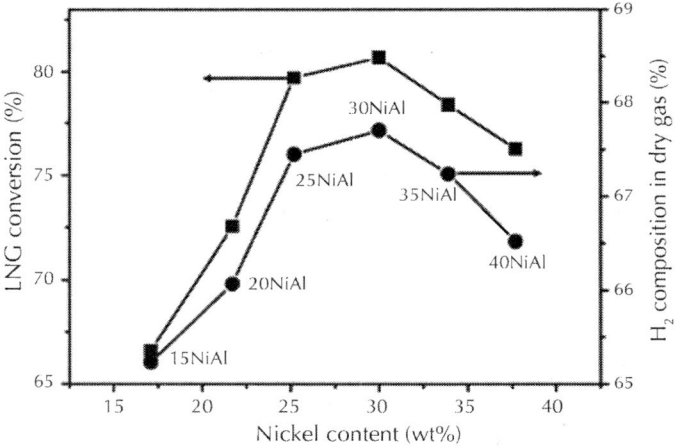

Figure 6: LNG conversions and H_2 compositions in dry gas over XNiAl (X = 15, 20, 25, 30, 35, and 40) catalysts in the steam reforming of LNG, plotted as a function of nickel content. The catalytic reaction data were obtained at 600 °C after a 400-min reaction. All catalysts were reduced at 700 °C prior to the reaction.

Correlations between Reducibility and Catalytic Activity

Although the reducibility of the catalyst is not the sole factor determining the catalytic performance in the steam reforming reaction, it can serve as a correlating parameter for the catalytic performance in the steam reforming reaction. Fig. 7 shows the correlations between reducibility (reduction peak temperature) and catalytic activity of XNiAl (X = 15, 20, 25, 30, 35, and 40) catalysts in the steam reforming of LNG. The reduction peak temperature determined by TPR measurements (Fig. 4) increased in the order of 30NiAl (763 °C) < 25NiAl (773 °C) < 35NiAl (790 °C) < 40NiAl (805 °C) < 20NiAl (821 °C) < 15NiAl (846 °C). The lower reduction peak temperature corresponds to the higher reducibility of the catalyst. It should be noted that both LNG conversion and H_2 composition in dry gas were well correlated with the reducibility of the catalyst. Both LNG conversion and H_2 composition in dry gas increased with increasing reducibility of the catalyst; more reducible catalyst showed a better catalytic performance in the steam reforming of LNG. It is believed that the surface area of metallic nickel was higher in the more reducible catalyst through the formation of surface nickel aluminate phase. Among the catalysts tested, the 30NiAl catalyst with the highest reducibility showed the best catalytic performance. Therefore, it is concluded that optimal nickel content was required for maximum catalytic performance of mesoporous nickel–alumina xerogel catalysts in the steam reforming of LNG.

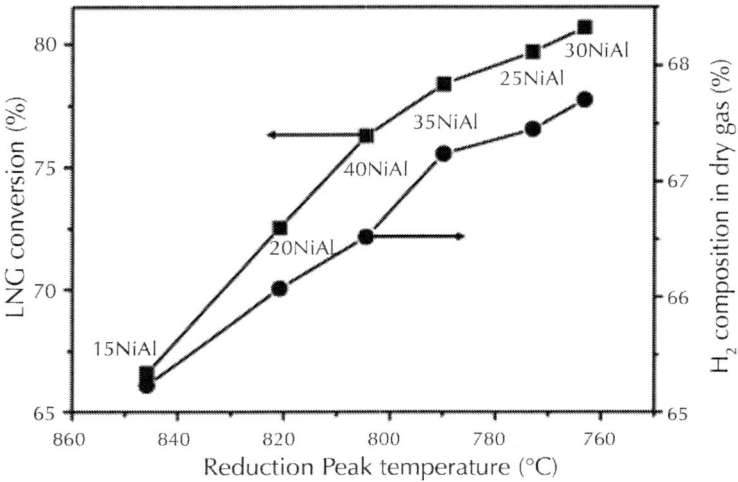

Figure 7: Correlations between reducibility and catalytic activity of XNiAl (X = 15, 20, 25, 30, 35, and 40) catalysts in the steam reforming of LNG. The catalytic reaction data were obtained at 600 °C after a 400-min reaction.

CONCLUSIONS

A series of mesoporous nickel–alumina xerogel catalysts with various nickel contents were prepared by a single-step sol–gel method, and were applied to hydrogen production by steam reforming of LNG. The effect of nickel content on the catalytic performance of XNiAl (X = 15, 20, 25, 30, 35, and 40) catalysts was investigated. Physical properties of XNiAl (X = 15, 20, 25, 30, 35, and 40) catalysts were affected by nickel content. Nickel species were finely dispersed in the XNiAl (X = 15, 20, 25, 30, 35, and 40) catalysts through the formation of Ni–O–Al composite structure. The XNiAl (X = 15, 20, 25, 30, 35, and 40) catalysts showed a stable catalytic performance in the steam reforming of LNG. It was found that both LNG conversion and H_2 composition in dry gas showed volcano-shaped curves with respect to nickel content, and decreased in the order of 30NiAl > 25NiAl > 35NiAl > 40NiAl > 20NiAl > 15NiAl.

It was also revealed that both LNG conversion and H_2 composition in dry gas increased with increasing reducibility of the catalyst. Among the catalysts tested, the 30NiAl catalyst with the highest reducibility showed the best catalytic performance. The highest surface area and the largest pore volume of the 30NiAl catalyst also played important roles in enhancing the gasification reaction between adsorbed hydrocarbons and steam. In conclusion, the mesoporous nickel–alumina xerogel catalysts prepared by a single-step sol–gel method served as efficient catalysts in the hydrogen production by steam reforming of LNG. In addition, optimal nickel content was required for maximum catalytic performance in the steam reforming of LNG.

ACKNOWLEDGMENTS

The authors wish to acknowledge support from the Seoul Renewable Energy Research Consortium (Seoul R & BD Program) and RCECS (Research Center for Energy Conversion and Storage: R11-2002-102-00000-0).

REFERENCES

1. M. Schrope, Nature 414 (2001) 682–684.
2. C.D. Blasi, G. Sinorelli, G. Portoricco, Ind. Eng. Chem. Res. 38 (1999) 2571–2581.
3. J.R. Rostrup-Nielsen, J. Sehested, J.K. Nørskov, Adv Catal. 47 (2002) 65–139.
4. J.N. Armor, Appl. Catal. A 176 (1999) 159–176.
5. A. Ishihara, E.W. Qian, I.N. Finahari, I.P. Sutrisna, T. Kabe, Fuel 84 (2005) 1462–1468.
6. M. Nurunnabi, Y. Mukinakano, S. Kado, B. Li, K. Kunimori, K. Suzuki, K. Fujimoto, K. Tomishige, Appl. Catal. A 299 (2006) 145–156.
7. J.K. Lee, D. Park, Korean J. Chem. Eng. 15 (1998) 658–662.

8. K.D. Ko, J.K. Lee, D. Park, S.H. Shin, Korean J. Chem. Eng. 12 (1995) 478–480.

9. J.R. Rostrup-Nielsen, Catal. Today 63 (2000) 159–164.

10. J. Sehested, J.A.P. Gelten, I.N. Remediakis, H. Bengaard, J.K. Nørskov, J. Catal. 223 (2004) 432–443.

11. S.W. Nahm, S.P. Youn, H.Y. Ha, S.-A. Hong, A.P. Maganyuk, Korean J. Chem. Eng. 17 (2000) 288–291.

12. A. Louloudi, N. Papayannakos, Appl. Catal. A 204 (2000) 167–176.

13. V. Tsipouriari, Z. Zhang, X.E. Verykios, J. Catal. 179 (1998) 283–291.

14. R. Molina, G. Poncelet, J. Catal. 199 (2001) 162–170.

15. Z. Pan, M. Dong, X. Meng, X. Zhang, X. Mu, B. Zong, Chem. Eng. Sci. 62 (2007) 2712–2717.

16. S. Cavallaro, V. Childo, A. Vita, S. Freni, J. Power Sources 123 (2003) 10–16.

17. L. Kepi ̦ nski, B. Stasi ́ nska, T. Borowiecke, Carbon 38 (2000) 1845–1856. ́

18. O. Yokota, Y. Oku, T. Sano, N. Hasegawa, J. Matsunami, M. Tsuji, Y. Tamaura, Int. J. Hydrogen Energy 25 (2000) 81–86.

19. K.O. Christensen, D. Chen, R. Lødeng, A. Holmen, Appl. Catal. A 314 (2006) 9–22.

20. T. Borowiecki, A. Gołebiowski, B. Stasi ̦ nska, Appl. Catal. A 153 (1997) ́ 141–156.

21. J.G. Seo, M.H. Youn, I.K. Song, J. Mol. Catal. A 268 (2007) 9–14.

22. A.C.S.C. Teixeira, R. Giudici, Chem. Eng. Sci. 56 (2001) 789–798.

23. K. Kochloefl, in: G. Ertl, H. Knozinger, J. Weitkamp (Eds.), Handbook of ¨ Heterogeneous Catalysis, vol. 4, Wiley, New York, 1997, pp. 1819–1831.

24. P. Kim, Y. Kim, H. Kim, I.K. Song, J. Yi, Appl. Catal. A 272 (2004) 157–166.

25. E.D. Dimotakis, T.J. Pinnavaia, Inorg. Chem. 29 (1990) 2393–2394.

26. T. Borowiecki, G. Wojciech, D. Andrzej, Appl. Catal. A 270 (2004) 27– 36.

27. J.S. Lisboa, D.C.R.M. Santos, F.B. Passos, F.B. Noronha, Catal. Today 101 (2005) 15–21.

28. S. Natesakhawat, R.B. Watson, X. Wang, U.S. Ozkan, J. Catal. 234 (2005) 496–508.

29. J.-H. Kim, D.J. Suh, T.-J. Park, K.-L. Kim, Appl. Catal. A 197 (2000) 191–200.

30. D.J. Suh, T.-J. Park, J.-H. Kim, K.-L. Kim, J. Non-Cryst. Solids 225 (1998) 168–172.

31. J.G. Seo, M.H. Youn, K.M. Cho, S. Park, I.K. Song, J. Power Sources 173 (2007) 943–949.

32. T. Ueckert, R. Lamber, N.I. Jaeger, U. Schubert, Appl. Catal. A 155 (1997) 75–85.

33. G. Li, L. Hu, J.M. Hill, Appl. Catal. A 301 (2006) 16–24.

34. M.L. Jacono, M. Schiavello, A. Cimino, J. Phys. Chem. 75 (1971) 1044–1050.

35. A.N. Kharat, P. Pendleton, A. Badalyan, M. Abedini, M.M. Amini, J. Catal. 205 (2002) 7–15.

36. M. Wu, D.M. Hercules, J. Phys. Chem. 83 (1979) 2003–2008.

Thermoeconomic Modeling and Exergy Analysis of a Decentralized Liquefied Natural Gas-fueled Combined-cooling–heating-and-power Plant

Alexandros Arsalis and Andreas Alexandrou

Research Center for Sustainable Energy, Department of Mechanical and Manufacturing Engineering, University of Cyprus, Nicosia, Cyprus

ABSTRACT

A small-scale combined-cooling–heating-and-power (CCHP) plant is proposed as a possible alternative to large-scale, centralized, electricity-only power plants. The proposed system is based on a non-ideal gas turbine (Brayton) cycle, integrated with a cooling plant and a district energy network. The study analyzes whether the proposed system could be an ideal candidate for distributed generation applications, especially in locations which are distant from centralized power plants. Therefore apart from reducing transmission and distribution losses, waste heat could be recovered effectively to generate heating, or cooling (via a cooling plant based on absorption refrigeration technology). The system considers fueling with liquefied natural gas (LNG), which is a safe and transportable fuel option. The cooling energy in the LNG is recovered in a useful manner since LNG regasification is coupled to the ambient air feed to the air compressor. The study includes a basic thermodynamic analysis, followed by an exergy analysis and a cost analysis.

The simulation results signify a potential for further investigation of the proposed system, since its performance results in significant thermodynamic and environmental improvements, when compared to an equivalent conventional system. The system operates in two modes: (a) winter operation, where recovered heat is distributed to the district energy network, (b) summer operation, where recovered heat is used to drive an absorption chiller cooling plant to generate cooling, which is also distributed to the district energy network. The average primary energy ratio of the proposed system is 0.91, while the net electrical efficiency is 0.356 and 0.365, for summer and winter operation, respectively. The average exergetic efficiency is 0.419, which is a 25% improvement when compared to the conventional system. The cost analysis shows that the payback period is within a reasonable time frame (approximately 4 years), for a total initial cost of 17.1 million .

INTRODUCTION

Combined-heat-and-power (CHP) is a well-established technology that utilizes the waste heat generated from an electricity-generating engine. In later years this technology has expanded in the additional generation of cooling, which is typically generated in an absorption refrigeration cycle (Arsalis, 2012). In such a combined-cooling–heating-and-power (CCHP) system, heat is used to feed the generator of a closed absorption cycle to generate cooling. This generated cooling energy (usually in the form of cold water) can be used for refrigeration or air conditioning (space cooling) purposes, depending on the type of the working fluid of the refrigeration cycle (Herold et al., 1996). In the case of space cooling, the most convenient working fluid combination is LiBr–water, due to the working level of the cycle, which is able to produce efficiently low quality cooling, i.e. cooling at 5–20 °C. The COP is very high for this cycle and no rectifier is necessary, as in the case of water–NH$_3$ based cycles (Herold et al., 1996).

A common CCHP system design may include a gas turbine, where electricity is generated in a non-ideal Brayton cycle and waste heat is recovered by means of heat exchangers. Therefore such a system is able to utilize the chemical energy of the fuel up to 90%, instead of 30–40%, which is the net electrical efficiency of the gas turbine cycle. Such a configuration allows the reduction in electricity and fuel consumption, since waste heat can be recovered in a useful manner, i.e. for hot water generation, space heating and space cooling applications. This translates to reduction of fuel consumption when compared to separate production of heat and power, e.g. a boiler would be needed to generate heat, or reduction of electricity consumption in the case of vapor-compression (electric) heat pumps used to generate space heating, space cooling and/or hot water.

Literature Review

Numerous combinations of CCHP systems have been proposed and are available in the literature. These studies consider the thermodynamic characteristics of the proposed systems, with associated cost/feasibility analysis. Kong et al. (2004) considered a 50 kWe Stirling engine system coupled to an absorption chiller. The system was analyzed and assessed with the inclusion of parameters such as the primary energy rate and comparative primary energy saving, while the cost model included factors such as the avoided cost, the total annual saving and the paying period. It was also compared to separate generation of heat, cool and power to investigate possible advantages in terms of fuel reduction. The recovered heat from the system was 146 kW and the payback period (PP) was 3.4 years. Wang et al. (2011) investigated the operational characteristics of a CCHP system for different scenarios, i.e. electricity-led and heat-led operation, to calculate thermodynamic and environmental parameters of interest. The system was compared to separate production of heat, cooling and power systems, in terms of primary energy savings, exergy efficiency and CO_2 emissions reduction. It was concluded that the performance of the proposed system depends strictly on demand.

Moné et al. (2001) investigated the economic feasibility of a CCHP system based on a gas turbine coupled to an absorption chiller. All available types of chiller were considered in terms of effect (single, double, and triple). The size, exhaust temperature and exhaust flow rate of the gas turbine are used to determine the maximum heat input rate to the chiller. It was calculated that a 250 MWe gas turbine can generate 150–300 MW of cooling, depending on the number of effects for the chiller. The cost analysis concluded that important savings are possible, if gas turbine-based CCHP systems were to be adopted. Finally, Popli et al. (2012) investigated a gas turbine-based CCHP system, with waste heat recovered to generate steam. The steam is partly fed to a double-effect LiBr–water absorption chiller system, to reduce the temperature of the air feed to the air compressor. The remaining steam is used to partly meet the furnace heating load and supplement plant electrical power in a combined regenerative Rankine cycle. The analysis concluded that the system can recover 79.7 MW of waste heat (37.1 MW fed to the absorption chiller system and converted to 45 MW of cooling at 5 °C), while the steam turbine could generate 22.6 MWe. The annual savings, in terms of running costs, were US$ 20.9 million with a PP of 1 year.

The use of liquefied natural gas (LNG) as a fuel option has received significant attention not only because it is safe and easy to transport, but also due to the lower cost of liquefaction achieved in recent years (Gupta, 2012). LNG can be regasified in cryogenic power cycles to produce power, when large flow rates of LNG need to be regasified (Morosuk and Tsatsaronis, 2011). Alternatively LNG cooling energy can be recovered to cool feed air before compression in a gas turbine cycle (Morosuk and Tsatsaronis, 2011), or it can be utilized in a condenser to convert waste steam to water in a steam turbine cycle (Liu et al., 2009 and Wang et al., 2013). Shi et al. (2010) considered coupling a conventional combined cycle power plant with large-scale LNG regasification (delivered to the city-gas pipeline) by utilizing the LNG cold energy in both the condenser stage of the steam turbine cycle and as a cryogenic power cycle. The results showed that net electrical efficiency improved by 2.8%. Also 0.9 MWe less power was needed, since water pumps previously utilized in an LNG sea water regasification process were eliminated in the proposed system.

Research Objectives

The purpose of this research work is to investigate the feasibility potential of adopting a small-scale (12.826 MWe) LNG-fueled CCHP system, as an alternative to conventional large-scale power plants. Conventional power generation is assumed to be electricity-only in centralized locations, whereas cooling and heating is produced by vapor-compression heat pumps located in the serviced buildings. The proposed system consists of a commercially available gas turbine cycle coupled to three heat exchangers, a double-effect LiBr–water absorption chiller system and a district energy network, which delivers heating and cooling energy to nearby buildings. A gas turbine cycle is chosen over a steam turbine (Rankine) cycle, because the latter performs in significantly lower efficiencies. Additionally the utilization of waste heat in a steam turbine cycle is more difficult and more importantly it is available in lower temperature grades (e.g. condenser), which would make its recovery very inefficient for the absorption refrigeration cycle (only a single-effect absorption chiller could be coupled in this configuration). The research work also includes a cost analysis and an exergy analysis. The latter is adopted to investigate irreversibility factors, not evident by the basic (first law) thermodynamic analysis.

SYSTEM CONFIGURATION

The system configuration is shown in Fig. 1. Natural gas in liquefied form is transported to the power plant location and stored in tanks. By means of an LNG pump, fuel is pumped through a heat exchanger (HEx1), where it is heated from −160 °C to 10 °C, by means of ambient air, which in turn is cooled from ambient conditions and fed to the air compressor, where it is compressed before combustion. The high temperature flue gas generated in the combustor is used to drive the gas turbine and generate electricity by driving an electric generator. The exhausted flue gas is recovered by means of heat exchangers. During winter operation, heat is recovered in HEx3 to generate hot water, which is circulated through the district energy network to nearby households to provide space heating and hot water. During summer operation, heat is recovered in HEx2 to generate steam, which is provided to a double-effect LiBr–water absorption chiller to generate cooling, which similarly to winter operation, is distributed to buildings through the district energy network, for space cooling purposes.

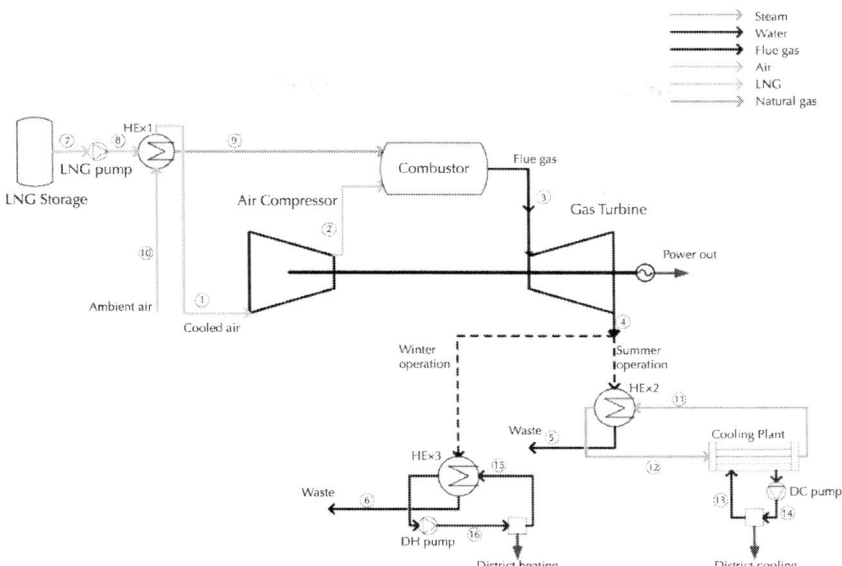

Figure 1: Schematic representation of the proposed combined cooling, heating and power system.

The system model includes the following characteristics and assumptions:

- Heat losses for every heat exchanger are assumed to be 5% for each side (i.e. hot and cold side) (Kakaç and Liu, 2002). Heat losses in the district energy network are assumed to be at an average value of 8%, which is typical for insulated tubes buried underground for a modern small-scale district energy network (Beith, 2011).

- Natural gas is assumed to be 100% methane for the purpose of calculation simplicity.

- Pressure drop in every heat exchanger (per side), pump, and combustor is assumed to be at a rate of 5% per component. Pressure drop in the district energy network is assumed to be at an average value of 10%.

- The power input to the pumps is considered negligible.

- The duration of each operational mode, summer or winter, is assumed to be six months.

- The proposed CCHP system is compared to an electricity-only generating, centralized power plant, which covers an equal load profile. Space heating, hot water and space cooling loads are fulfilled with vapor-compression heat pumps of an average COP-value of 3.0.

MODELING METHODOLOGY

The system is modeled based on theoretical and experimental (e.g. manufacturers' data) assumptions. Each component (or subsystem) is modeled separately and coupled to every adjacent component, as described in the System configuration. All details, including specific assumptions are given in the following subsections. The values for all system input parameters are given, followed by a description of all subsystem/component models, i.e. power plant, heat exchangers, cooling plant and district energy network. Finally the cost model is described in detail, with all necessary cost functions and their input values.

System Input Parameters

The system input parameters are taken from the literature or other sources to correspond to the operational characteristics and requirements of the system model. Their values are given in Table 1, along with the adopted notation, description and unit. The value of ambient temperature (T_{amb}) varies depending on season. The value of generator efficiency ($_{gen}$) is taken from Siemens (2013). The flue gas temperature exiting the system ($T_{fg,exh}$) for summer and winter operation is 145 and 65 °C, respectively, to allow proper operation of HEx2 and HEx3. A typical value of −160 °C is assumed for the LNG storage temperature (T_{LNG}) (Gupta, 2012), while the temperature of natural gas after regasification (T_{NG}) is 10 °C. The temperature of steam generated and fed to the absorption chiller ($T_{steam,}$s) is 150 °C and the return temperature ($T_{steam,}$s) is 142 °C (Herold et al , 1996). The cold water generated in the chiller is at a temperature ($T_{cw,}$s) of 7 °C, with a return temperature ($T_{cw,}$r) of 12 °C (Dinçer and Zamfirescu, 2011), while the coefficient of performance (COP_{th}) for the chiller is 1.3 (Herold et al., 1996). For winter operation, supply temperature ($T_{hw,}$s) is 80 °C and return temperature ($T_{hw,}$r) is 60 °C (Dinçer and Zamfirescu, 2011). The values for the pressure ratio (PR), isentropic efficiencies for compressor ($_{co}$) and turbine (t), and turbine exhaust temperature ($T_{gt,exh}$) are taken from Siemens (2013).

Table 1: Values of system model input parameters

Parameter Description		Value
T_{amb}	Ambient temperature	30/15 °C
P_{amb}	Ambient pressure	1 atm
$T_{fg,exh}$	Flue gas exhaust temperature (to ambient)	145/65 °C
T_{LNG}	LNG storage temperature	−160 °C
T_{NG}	Natural gas temperature (after regasification)	10 °C
$T_{steam,}$s	Steam supply temperature	150 °C
$T_{steam,}$r	Steam return temperature	142 °C
$T_{cw,}$s	Cold water supply temperature (summer)	7 °C
$T_{cw,}$r	Cold water return temperature (summer)	15 °C
$T_{hw,}$s	Hot water supply temperature (winter)	80 °C
$T_{hw,}$r	Hot water return temperature (winter)	60 °C
$T_{gt,exh}$	Turbine exhaust temperature (°C)	555
COP_{th}	Coefficient of performance (–)	1.3
PR	Compressor pressure ratio (–)	16.8
η_{gen}	Generator efficiency	0.972
η_{co}	Compressor isentropic efficiency (–)	0.831
η_t	Turbine isentropic efficiency (–)	0.901

Note: When two values are given, the first value is valid for summer operation, while the second one is valid for winter operation.

Component Models

The power plant model is based on a non-ideal gas turbine cycle (Klein and Nellis, 2012) consisting of a turbine, a compressor and a combustor. Application of this cycle in hot climates, with prolong periods of high temperatures, favors cooling of the ambient air stream before compression. Therefore the cooling energy from the LNG regasification process can be used to cool the air supply (Gupta, 2012). This heat transfer increases the specific volume of the air supply, while LNG is converted to natural gas without any external heat supply. This results to a lower compressor power input requirement, and thereby to a higher net electrical efficiency.

The specific enthalpy at the exit for the compressor is

$$\overline{h}_2 = \overline{h}_1 + \frac{\overline{h}_{s,2} - \overline{h}_1}{\eta_c} \tag{1}$$

Where $\overline{h}_1$ is the specific enthalpy at the inlet and $\overline{h}_{s,2}$ is the isentropic specific enthalpy at the exit.

The energy balance between reactants and products for the combustor is defined as follows:

$$\overline{h}_{CH_4,re} + 0.21a\,\overline{h}_{O_2,re} + 0.79a\,\overline{h}_{N_2,re}$$
$$= b\,\overline{h}_{CO_2,pr} + c\,\overline{h}_{H_2O,pr} + d\,\overline{h}_{N_2,pr} + e\,\overline{h}_{O_2,pr} \tag{2}$$

Where $\overline{h}_{i,re}$ and $\overline{h}_{i,pr}$ are the specific enthalpies of species i (on a mole basis) for reactants and products (state 3 in Fig. 1), respectively, and a–e are the reaction coefficients.

The net electrical power output is defined as

$$\dot{W}_{gen} = \dot{n}_{CH_4}\left[(H_{pr} - H_4) - a\left(\overline{h}_2 - \overline{h}_1\right)\right]\eta_{gen} \tag{3}$$

Where $\dot{n}_{CH_4}$ the molar flow rate of methane, H_{pr} is the enthalpy of the products exiting the combustor per mole of fuel, and H_4 is the enthalpy of the exhaust flue gas. Net electrical efficiency is defined as the ratio of the net electrical power output to the chemical energy of methane (on an LHV basis) as follows:

$$\eta_{el,net} = \frac{\dot{W}_{gen}}{\dot{n}_{CH_4} LHV} \tag{4}$$

The three HEx units are modeled with heat transfer balances as follows:

$$\dot{Q} = \dot{m}_c \left(h_{c,o} - h_{c,i} \right) = \dot{m}_h \left(h_{h,i} - h_{h,o} \right) \tag{5}$$

Where $\dot{m}_c$, $\dot{m}_h$ are the mass flow rates of the cold and hot side, respectively, while $h_{c,i}$, $h_{c,o}$ are the specific enthalpies of the cold side, inlet and outlet, respectively, and $h_{h,i}$, $h_{h,o}$ are the specific enthalpies of the hot side, inlet and outlet, respectively.

A double-effect LiBr–water absorption chiller can perform at COP-values in the range of 1.2–1.3 (Herold et al., 1996). The high quality of recovered heat could also allow the coupling of triple-effect absorption chillers, if the technology becomes commercialized at the capacity range of the proposed system (Deng et al., 2011). The COP_{th} value is defined as the ratio of the heat transfer rate of the evaporator to the heat transfer rate of the generator

$$COP_{th} = \frac{\dot{Q}_{evap}^{ach}}{\dot{Q}_{gen}^{ach}} \tag{6}$$

Cost Model

To allow a realistic analysis of the feasibility potential of the proposed system a rigorous cost model is formulated. The main economic parameters evaluated are: projected total lifecycle cost (LCC), investment cost, operating cost and PP. All cost functions are tabulated in Table 2, while all values of the input cost factors are given in Table 3 and all input values for the cost mode are given in Table 4. The latter values are taken fromDinçer and Zamfirescu (2011), unless specified otherwise.[1] The specific cost of LNG fuel (first year) is assumed to be at a value of 8.4 €/GJ, which is currently the average value in the European LNG market (Federal Energy Regulatory Commission, 2013). The lifetime for the power plant, cooling plant and district energy network is fixed at 25 years (Beith, 2011 and Dinçer and Zamfirescu, 2011). The specific cost of LNG transport is 0.11 €/ton-km (ESN-SECA, 2013), for an assumed LNG transport distance of 100 km.

Table 2: Cost model for the proposed CCHP system

Variable description (unit)		Model equation
E_{ey}	Annual electrical energy production (J/yr)	$E_{py} = \dfrac{E_{ey}}{\eta_{el,net}}$
E_{py}	Annual consumption of natural gas (J/yr)	
c_{fyy}	Annual cost of fuel (excl. regasification) (€/yr)	$c_{fyy} = E_{py}c_{LNG}$
c_{reg}	Annual regasification cost (€/yr)	$c_{rgf} = E_{py}c_{rgf}$
c_{fy}	Annual cost of fuel (incl. regasification) (€/yr)	$c_{fy} = c_{fyy} + c_{reg}$
$Q_{0}y$	Annual amount of rejected heat (J/yr)	$Q_{0y} = (1-\eta_{el,net})E_{py}$
Q_{hy}	Recovered heat (J/yr)	$Q_{hy} = (1-f)Q_{0y}$
f	Heat loss factor (–)	
Q_{cy}	Annual cooling energy production (J/yr)	$Q_{cy} = COP_{th}\dfrac{Q_{hy}}{2}$
$\dot{Q}_{c}$	Total chiller capacity (MW)	$\dot{Q}_c = \dfrac{Q_{cy}}{0.5\times365\times24\times3600}$
C_{cc}	Cost of cooling plant (€)	$C_{cc} = c_c\dot{Q}_c$
N_{fc}	Number of fan-coil units (–)	$N_{fc} = \dfrac{Q_{cy}}{Q_{fcy}}$

C_{dn}	Cost of distribution network (€)	$C_{dn} = Lc_l$
C_{pp}	Cost of power plant (€)	$$C_{pp} = c_p \dot{W}_{gen}$$
C_{fct}	Total cost of fan-coil units (€)	$$C_{fct} = c_{fc} N_{fc} \dot{Q}_{fc}$$
C_{cp}	Total cost of central plant (€)	$C_{cp} = C_{pp} + C_{cc}$
C_{CHP}	Total CHP cost (€)	$C_{CHP} = C_{cp} + C_{dn} + C_{fct}$
C_{down}	Down payment (€)	$C_{down} = (1 - f_{loan}) C_{CHP}$
AP_n	Capital recovery factor (–)	$$AP_n = \frac{r_n}{1 - (1 + r_n)^{-N}}$$
PA_n	Uniform series present worth factor (–)	$nP_A = (AP_n)^{-1}$
FP_n	Compound amount factor (–)	$nFP = (1 + rn)^{-N}$
PF_n	Present worth factor (–)	$PF_n = (FP_n)^{-1}$
C_{loan}	Cost of the loan (€)	$$C_{loan} = \frac{AP_1}{AP_2} f_{loan} C_{CHP}$$
D_{loan}	Tax deduction on the loan (€)	$$D_{loan} = t f_{loan} C_{CHP} \left(\frac{AP_1}{AP_2} - \frac{AP_1 - r_1}{(1 + r_1) AP_3} \right)$$
C_{twc}	Total worth of capital (€)	$C_{twc} = C_{down} + C_{loan} - D_{loan}$
D_{dep}	Linear depreciation of capital (€)	$$D_{dep} = \frac{C_{CHP}}{N} t\, PA_2$$
D_{cred}	Tax credit (€)	$D_{cred} = t_{cred} C_{CHP}$
D_{salv}	Salvage worth (€)	$D_{salv} = f_{salv} C_{CHP} PF_2 (1 - t_{salv})$
C_{prop}	Tax paid on property (€)	$C_{prop} = f_{prop} C_{CHP} t_{prop} (1 - t)$
C_{omi}	Operation, maintenance and insurance cost (€)	$$C_{omi} = f_{omi} C_{CHP} \frac{PA_2}{3} (1 - t)$$
C_{tcf}	Total cost of fuel (€)	$$C_{tcf} = c_{fy} \left(\frac{1 - t}{AP_4} \right)$$

C_{tr}	Total cost of LNG transport (€)	$C_{tr} = c_{tr} L_{tr} V_{truck} s p_{LNG,yr} N$
C_{life}	Lifecycle cost (€)	$C_{life} = C_{tr} + C_{twc} + C_{prop} + C_{omi} + C_{tcf} - (D_{dep} + D_{cred} + D_{salv})$
PP	Payback period (yr)	$$PP = \dfrac{C_{CHP}}{PES\, c_{fy}}$$

Table 3: Input cost factor values

Parameter description (unit)		Value
AP_1	Capital recovery factor 1 (–)	0.064
AP_2	Capital recovery factor 2 (–)	0.071
AP_3	Capital recovery factor 3 (–)	0.202
AP_4	Capital recovery factor 4 (–)	0.021
PA_2	Uniform series present worth factor (–)	14.09
PF_2	Present worth factor (–)	3.386
r_1	Rate factor 1 (–)	0.04
r_2	Rate factor 2 (–)	0.05
r_3	Rate factor 3 (–)	0.2
r_4	Rate factor 4 (–)	−0.045

Table 4: Input values for the cost model

Input Parameter Description		Value
t	Loan time factor	0.4
t_{cred}	Tax credit time factor	0.02
t_{prop}	Tax on property time factor	0.25
t_{salv}	Salvage worth time factor	0.2
f_{loan}	Loan factor	0.8
f_{omi}	Operation, maintenance and insurance factor	0.01
f_{prop}	Tax on property factor	0.5
f_{salv}	Salvage worth factor	0.1
N	Number of years of service	25

f	Heat loss factor	0.08
$\dot{Q}_{fc}$	Fan-coil energy input rate	0.5 kW
L	Length of pipe network	0.5 km
L_{tr}	LNG transport distance	100 km
V_{truck}	Volume per truck	58 m³
Q_{fcv}	Total energy of fan-coil units	16.1 GJ
lf	Load factor	1
c_{fc}	Specific cost of fan-coil unit	73 €/kW
cp	Specific cost of natural gas fired power plant	365 €/GW
cl	Specific cost of distribution line	2.08 €/km
cc	Specific cost of cooling plant	365 €/kW
c_{LNG}	Specific cost of LNG fuel for the first year	8.4 €/GJ
c_{rgf}	Specific cost of LNG regasification	0.28 €/GJ
c_{tr}	Specific cost of LNG transport	0.11 €/ton-km

EXERGY ANALYSIS

Exergy analysis provides insight on system irreversibilities, which are not evident from an analysis based only on the first law of thermodynamics. Exergy analysis (or second law) compares any state of the system to an equivalent "dead state", and therefore the potential of a particular state or process can be evaluated. A "dead state" is the state that allows no further exploitation of useful energy (equivalently no work can be produced), since all thermodynamic parameters (i.e. pressure, temperature, kinetic energy, potential energy) have reached equilibrium with the surroundings. Therefore exergy analysis is a useful tool in an effort to improve performance, by assessing thermodynamic irreversibilities in terms of magnitude, location and cause (Morosuk and Tsatsaronis, 2011). For the current research study, dead state conditions are assumed to be equal to those at atmospheric conditions (i.e. $T_0 = T_{amb}$, $P_0 = P_{amb}$).

The exergy rate of net electrical output is equal to the value of net power output

$$\dot{X}_{el} = \dot{W}_{gen} \tag{7}$$

The exergy rate of district heating during winter operation, or district cooling during summer operation, delivered by the energy network is defined as follows, respectively (Wang et al., 2011):

$$\dot{X}_h = \dot{Q}_{DHN}\left(1 - \frac{T_0}{T_{16}}\right) \tag{8}$$

$$\dot{X}_c = \dot{Q}_{DCN}\left(\frac{T_0}{T_{14}} - 1\right) \tag{9}$$

The exergy rate of fuel is defined as follows:

$$\dot{X}_{fuel} = \dot{m}_9\left(\frac{\overline{x}^{CH_4}_{CH_4}}{M_{w,CH_4}}\right) \tag{10}$$

Where $\overline{x}^{CH_4}_{CH_4}$ the standard molar chemical exergy of methane and M_w, CH_4 is the molecular weight of methane.

The second law efficiency (also termed as the "exergetic efficiency") is the sum of exergy rate output to the exergy of the fuel provided to the system, where $\dot{X}_{ch} = X_c$ for summer operation, and $\dot{X}_{ch} = X_h$ for winter operation.

$$\eta_2 = \frac{\dot{W}_{gen} + \dot{X}_{ch}}{\dot{X}_{fuel}} \tag{11}$$

The specific physical exergy for any state i can be defined as follows (Bejan et al., 1996):

$$x^{PH}_{f,i} = h_i - h_{0,i} - T_0\left(s_i - s_{0,i}\right) \tag{12}$$

Where h_i is the specific enthalpy at state i, h_0, i is the specific enthalpy at the dead state, s_i is the specific entropy at state i, and s_0, i is the specific entropy at the dead state.

Therefore the physical exergy rate for any state i can be defined as

$$\dot{X}_i^{PH} = \dot{m}_i x_{f,i}^{PH}$$

(13)

Similarly the chemical exergy rate of any state i can be defined as

$$\dot{X}_i^{CH} = \dot{m}_i x_{f,i}^{CH}$$

(14)

Where $x_{f,i}^{CH}$ is the specific chemical exergy for any state i with values taken from Bejan et al. (1996).

The total exergy rate for any state i is the sum of chemical exergy rate and physical exergy rate

$$\dot{X}_i = \dot{X}_i^{CH} + \dot{X}_i^{PH}$$

(15)

An exergy balance for the overall system can be written as follows:

$$\dot{X}_{total,i} = \dot{X}_{total,o}$$

(16)

The total exergy rate entering and exiting the system is defined as follows, respectively:

$$\dot{X}_{total,i} = \dot{X}_{10} + \dot{X}_7$$

(17)

$$\dot{X}_{total,o} = \dot{W}_{gen} + \dot{X}_{ch} + \dot{X}_{loss} + \dot{X}_{des}$$

(18)

Where $\dot{X}_{des}$ is the total exergy destruction rate it is defined as follows, for winter and summer operation, respectively:

$$\dot{X}_{des} = \dot{X}_{des,ac} + \dot{X}_{des,gt} + \dot{X}_{des,comb} + \dot{X}_{des,he1} + \dot{X}_{des,he3}$$

(19)

$$\dot{X}_{des} = \dot{X}_{des,ac} + \dot{X}_{des,gt} + \dot{X}_{des,comb} + \dot{X}_{des,he1} + \dot{X}_{des,he2} + \dot{X}_{des,cp}$$

(20)

Exergy balances for every system component can be formulated to allow calculation of the rates of exergy destruction (irreversibility) per component. These equations are tabulated in Table 5.

Table 5: Exergy balance equations for every system component

Component	Equation
Air compressor	$\dot{m}_1 x_{f,1} + \dot{W}_{ac} = \dot{m}_2 x_{f,2} + \dot{X}_{des,ac}$ where $\dot{W}_{ac} = \dot{m}_1(h_2 - h_1)$
Gas turbine	$\dot{m}_3 x_{f,3} = \dot{m}_4 x_{f,4} + \dot{W}_{gt} + \dot{X}_{des,gt}$ where $\dot{W}_{gt} = \dot{m}_3(h_3 - h_4)$
Combustor	$\dot{X}_{fuel} + \dot{m}_2 x_{f,2} + \dot{m}_9 x_{f,9} = \dot{m}_3 x_{f,3} + \dot{X}_{des,comb}$
Heat exchanger 1	$\dot{m}_8 x_{f,8} + \dot{m}_{10} x_{f,10} = \dot{m}_1 x_{f,1} + \dot{m}_9 x_{f,9} + \dot{X}_{des,he1}$
Heat exchanger 2	$\dot{m}_4 x_{f,4} + \dot{m}_{11} x_{f,11} = \dot{m}_5 x_{f,5} + \dot{m}_{12} x_{f,12} + \dot{X}_{des,he2}$
Heat exchanger 3	$\dot{m}_4 x_{f,4} + \dot{m}_{15} x_{f,15} = \dot{m}_6 x_{f,6} + \dot{m}_{16} x_{f,16} + \dot{X}_{des,he3}$
Cooling plant	$\dot{m}_{14} x_{f,14} + \dot{m}_{11} x_{f,11} = \dot{m}_{13} x_{f,13} + \dot{m}_{15} x_{f,15} + \dot{X}_{des,cp}$

The term $\dot{X}_{loss}$ is the exergy rate lost due to the inability of the system to utilize effectively the remaining exergy rate released to the surroundings. It is defined as follows, for winter and summer operation, respectively:

$$\dot{X}_{loss} = \dot{X}_6 \tag{21}$$

$$\dot{X}_{loss} = \dot{X}_5 \tag{22}$$

The exergy destruction ratio for any component k is the ratio of the component's exergy destruction rate to the total exergy rate input (Bejan et al., 1996)

$$y_{D,k} = \frac{\dot{X}_{des,k}}{\dot{X}_7} \tag{23}$$

The total rate of entropy generation can be calculated using the Gouy–Stodola theorem (Bejan et al., 1996)

$$\dot{S}_{gen} = \frac{\dot{X}_{des}}{T_0}$$

(24)

The exergy loss ratio is defined as the ratio of exergy rate loss to the total exergy rate input (Bejan et al., 1996)

$$y_L = \frac{\dot{X}_{loss}}{\dot{X}_7}$$

(25)

RESULTS AND DISCUSSION

Validation

The system model is validated using two commercially available gas turbines as reference cases (SGT-400 Industrial Gas Turbine (Siemens, 2013) and GE10-2 Gas Turbine (GE, 2006)), which include measured data from their respective manufacturer. Therefore the model input values correspond to the values taken from the references. The results of the model validation are given in Table 6. The calculated error between the system simulation and the manufacturers' data is low (0–4.4%). Thereby, it can be assumed that the proposed system model is modeled within an acceptable accuracy level.

Table 6: Validation of the simulated system model

Input Parameter Description	Reference 1[a]	Reference 2[b]
Power generation (MWe)	12.90	11.98
Compressor exit pressure (MPa)	1.702	1.571
Exhaust temperature (°C)	555	480

Performance Parameter Description	Reference 1	Simulation	Error (%)	Reference 2	Simulation	Error (%)
Electrical efficiency (-)	0.348	0.348	0	0.333	0.334	0.3
Exhaust gas flow (kg/s)	39.4	41.2	4.4	47.0	47.3	0.6

[a] SGT-400, (Siemens, 2013).

[b] GE10-2, (GE, 2006).

Thermodynamic Data

The system model calculates all unknown thermodynamic properties (e.g. temperature, pressure) for all state points, which are shown in Table 7. It is evident that the heat transfer in HEx1 results in a significant decrease of air feed temperature to the compressor, which is equal to 15 °C temperature drop. This modification results to a net electrical efficiency increase of almost 2%. The three pumps shown in the system configuration (see Fig. 1) are assumed to require negligible power input for different reasons. LNG pump pressurizes natural gas from 1 MPa to 4.5 MPa, but since the respective flow rate is too low, the power input requirement is not affecting the net power output. The other two pumps (DH pump and DC pump) circulate hot and cold water through the district energy network. They carry large flow rates, but since fluid flow is within closed circuits, only pressure losses can reduce the flow rate. They also flow at a marginal pressure gradient (0.10–0.12 MPa), as compared to atmospheric pressure, which is high enough to maintain circulation.

Table 7: Thermodynamic data for the proposed CCHP system of Fig. 1

State	Substance	Temperature (°C)	Pressure (MPa)	Mass flow rate (kg/s)	Chemical exergy rate (MW)	Physical exergy rate (MW)	Total exergy rate (MW)
1	Air	15/−1	0.101	37.84/35.42	0	0	0
2	Air	431/394	1.702	37.84/35.42	0	15.1	15.1
3	Flue gas	1179/1178	1.617	38.56/36.12	0.2/0.3	40.5/38.5	40.7/38.8
4	Flue gas	555	0.107	38.56/36.12	0.2/0.3	9.6/9.7	9.8/10.0
5	Flue gas	145/NA	0.102	38.56/NA	0.2/NA	0.7/NA	0.9/NA
6	Flue gas	NA/65	0.102	NA/36.12	NA/0.3	NA/0.2	NA/0.5
7	LNG	−160	1.000	0.72/0.70	36.9/36.0	0.8/0.7	37.7/36.7
8	LNG	−160	4.500	0.72/0.70	36.9/36.0	0.8/0.7	37.7/36.7
9	Natural gas	10	4.275	0.72/0.70	36.9/36.0	0.4/0.4	37.3/36.4
10	Air	30/15	0.101	37.84/35.42	0	0	0
11	Steam	142/NA	0.106	1105.93/NA	2.8/NA	529.6/NA	532.3/NA
12	Steam	150/NA	0.106	1105.93/NA	2.8/NA	534.4/NA	537.2/NA
13	Water	15/NA	0.109	682.53/NA	1.7/NA	1.1/NA	2.8/NA
14	Water	7/NA	0.122	682.53/NA	1.7/NA	2.6/NA	4.3/NA
15	Water	NA/60	0.109	NA/233.44	NA/0.6	NA/3.1	NA/3.7
16	Water	NA/80	0.122	NA/233.44	NA/0.6	NA/6.2	NA/6.8

Note: When two values are given, the first value represents conditions for summer operation and the second value represents conditions for winter operation.

The main performance characteristics of the proposed system are given in Table 8. The system provides more cooling (21 MW) than heating (18 MW), due to the utilization of a refrigeration cycle with a COP-value above unity. Also more heat is available for recovery during summer operation because the system performs more poorly during summer operation, mainly because of the higher ambient conditions. Therefore net electrical efficiency is lower during summer operation (35.6%), as compared to winter operation (36.5%). Also during summer operation the system cannot utilize heat as effectively as in winter operation, due to the operational requirements of the cooling plant. Specifically flue gas is exhausted at 65 and 145 °C, during winter and summer operation, respectively. Coupling of a second cooling plant, i.e. single-effect LiBr absorption chiller operating at lower generator heat input would allow further exploitation of waste heat. However the benefit would be marginal to the overall system performance and additional capital cost would be required to purchase the extra equipment. The respective LNG consumption per year is 53.1 tons. The practicality of adopting such a system must be considered in terms of fuel transport to the power plant site. If the power plant would be fueled with LNG transported through LNG trucks, 916 truck deliveries per year (assuming each truck can carry 58 m³ of LNG (Linde, 2011)) would be required. But it should be noted that these values are only valid for continuous full-load operation. The system would be more likely operated at an average load of 75%, in which case the fueling frequency would drop to less than 700 deliveries per year.

Table 8: Main performance characteristics of the proposed CCHP system

Performance parameter description		Valuea
Efuel,i	Fuel chemical energy consumption (MW)	35.9/35.1
$\dot{W}_{gen}$	Net electrical power output (MWe)	12.8
$\dot{Q}_{DCN}$	Cooling energy output (MW)	21.0

$\dot{Q}_{DHN}$	Heating energy output (MW)	18.0
el,net	Net electrical efficiency (–)	0.356,0.365
$\dot{V}_{LNG,yr}$	LNG volumetric flow rate (ton/yr)	53.1
spLNG,yr	LNG supplyb frequency (times/yr)	916

[a] When two values are given, the first value represents conditions for summer operation and the second value represents conditions for winter operation.

[b] LNG is assumed to be transported to the site of the power plant with LNG trucks (volume capacity per truck is 58 m³).

The advantages of the proposed trigeneration system can be showed with a direct comparison with an equivalent conventional system. The conventional system assumes an electricity-only gas turbine-based power plant (fueled with natural gas) and generation of electricity at a centralized location, while heating and cooling is generated in the location of the consumers with electric heat pumps. Therefore the conventional system must generate more electricity in order to fulfill the electricity input needs for the heat pumps. Also for the conventional system, 8% transmission and distribution losses (f_{lc} = 0.08) are assumed in the electricity grid. Table 9 shows four key performance indicators used to evaluate the effectiveness of the proposed system as compared to the equivalent conventional system. These parameters are: electricity savings

($\dot{W}_{el,net,sav}$ = 8.2MWe), CO_2 emissions reduction (ERCO$_2$=38.9%), primary energy savings (PES = 41.8%), and primary energy ratio (PER = 0.91). Clearly the savings in power consumption, reduction in CO_2 emission generation, and fuel consumption are very significant, and verify the potential of the system, as an alternative to separate heat and power generation.

Table 9: Performance comparison of the proposed CCHP system and an equivalent conventional system

Performance parameter	Definition	Average value
Electricity savings	$\dot{W}_{net,el,sav} = \frac{\dot{W}_{gen} + \frac{Q_{DE}}{COP_{conv}}}{1 - f_{ic}} - \dot{W}_{gen}$	8.2 MWe
CO$_2$ emissions reduction	$ER_{CO_2} = 1 - \frac{\dot{m}_{CO_2}}{\dot{m}_{CO_2,conv}} \times 100\%$	38.9%
Primary energy savings	$PES = 1 - \frac{E_{fuel,i}}{E_{fuel,i,conv}} \times 100\%$	41.8%
Primary energy ratio	$PER = \frac{\dot{W}_{gen} + \dot{Q}_{DE}}{E_{fuel,i}}$	0.91

Cost Analysis

Based on the cost model analyzed in subsection 3.3, the cost parameters are calculated and presented in Table 10. As shown from the results the total initial cost (C_{CHP}) is 17.1 million €. The cost of the power plant is 4.7 million €, which is less than half when compared to the cost of the cooling plant (10.1 million €). This is due to the high specific cost of the cooling plant, because absorption chiller technology is still an emerging technology and not as mature as gas turbine technology. However the economic trend of absorption refrigeration in the future is expected to be available at lower costs (Deng et al., 2011), which will allow lower overall costs than the one calculated here. The total lifecycle cost (C_{life}) is 292.4 million €. It is obvious that LCC strongly depends on operational costs, while capital (investment) costs have a lesser significance. This means that the LCC will be strongly depended on the fluctuation of LNG price throughout the lifetime of the CCHP system. Therefore the calculated LCC value can only provide an indication of the expected cost, within a high degree of uncertainty.

Table 10: Cost analysis for the proposed CCHP system

Variable description (unit)		Value (million)
C_{cc}	Cost of cooling plant ( )	10.1
C_{dn}	Cost of distribution network ( )	1.3
C_{pp}	Cost of power plant ( )	4.7
C_{fct}	Total cost of fan-coil units ( )	1.0
C_{cp}	Total cost of central plant ( )	14.7
C_{CHP}	Total CHP cost ( )	17.1
C_{down}	Down payment ( )	3.4
C_{loan}	Cost of the loan ( )	12.3
C_{twc}	Total worth of capital ( )	11.4
C_{prop}	Tax paid on property ( )	1.3
C_{omi}	Operation, maintenance and insurance cost ( )	1.4
C_{tcf}	Total cost of fuel ( )	285.1
C_{tr}	Total cost of LNG transport ( )	5.1
C_{life}	Lifecycle cost ( )	292.4

The variation of LCC and PP for different values of LNG cost is shown in Fig. 2. The LCC varies linearly due to its high dependence on the LNG cost. For cost prices above 12 /GJ, the LCC exceeds 400 million , which would not allow a high potential for the proposed system to enter the energy market, especially if other cheaper fuel-driven alternatives are available. On the other hand, from a PP perspective, a high LNG cost results to a lower PP, when the comparison is in relation to the conventional system. This is due to the more effective utilization of the chemical energy of the fuel in the case of the proposed CCHP system. Therefore a very low LNG price would result in a very high PP value for the proposed system, and the investment potential would become very unlikely due to the high risk for investment. Apart from the system design point (i.e. 12.8 MWe), the system is expected to operate within a range from 50% load to full load. Fig. 3 shows the fluctuation of the LCC and PP values for different load factor values. The figure trend lines suggest that if the system is operated close to a 75–100% load, the PP value will only increase slightly, as opposed to continuous operation at full load. However if the system is operated at a load factor value close to 50%, the PP value would increase by approximately 1.5 years, which in addition to lower efficiency performance at these loads, the proposed system would have limited chances of penetrating the energy market.

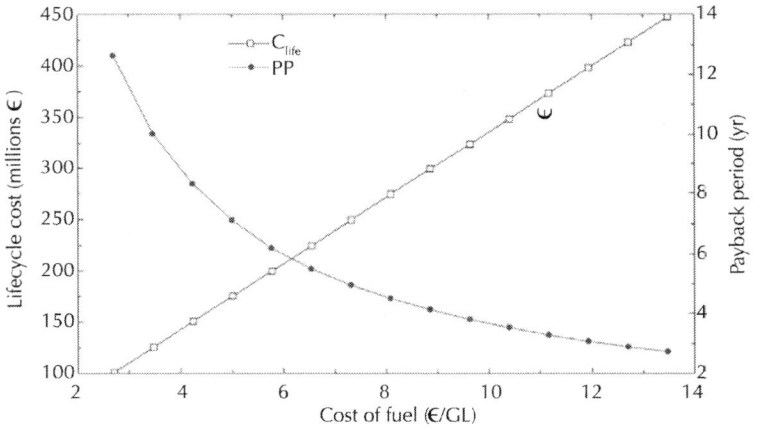

Figure 2: Variations in payback period and lifecycle cost for different fuel price values.

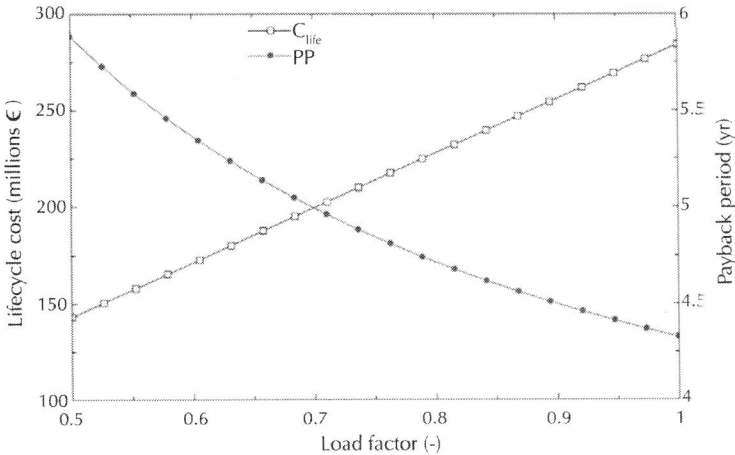

Figure 3: Variations in payback period and lifecycle cost for different load factor values.

Part-load Operation

The performance of the gas turbine cycle at difference loads is shown graphically in Fig. 4. This turbine map is constructed with data taken from Siemens (2013), and it is useful for determining performance at part-load operation. This is performed with two-dimensional interpolation, where the function inputs are the ambient temperature and the net electrical power output of the system, while the function outputs are the heat input rate and the turbine exhaust temperature. The net electrical efficiency at part-load is calculated as the ratio of net electrical power output to heat input rate

$$\eta_{el.net}^{pl} = \frac{\dot{W}_{el.net}}{\dot{Q}_i} \qquad (26)$$

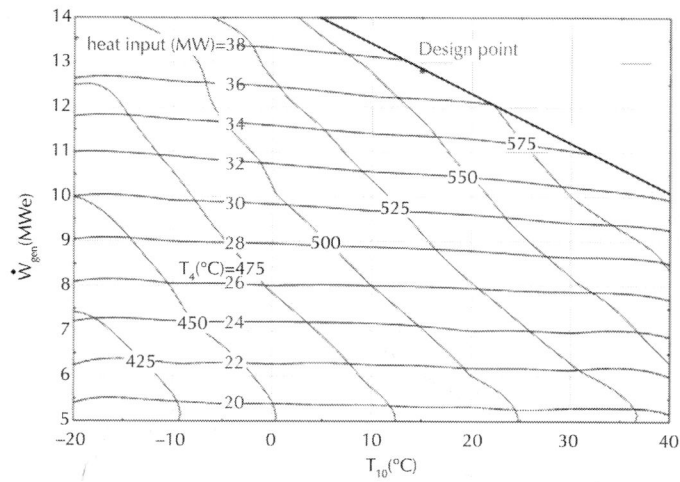

Figure 4: Map of the modified gas turbine cycle.

However, to acknowledge the fact that the gas turbine cycle of the proposed system includes air cooling of the air feed provided to the compressor, a correction factor (cf_{gt}) is multiplied to the result of the previous equation, to give a corrected net electrical efficiency value, as follows:

$$\eta_{el,net}^{pl,cf} = \eta_{el,net}^{pl} cf_{gt} \tag{27}$$

The cf_{gt} values are 1.0667 and 1.0580, for winter and summer operation, respectively. These values are given based on the simulation results at full-load operation. For example, winter operation at 50 and 75% load would result in a net electrical efficiency of 0.304 and 0.343, respectively, while summer operation at the same loads would result in an efficiency of 0.299 and 0.335, respectively.

Application of the Proposed System

The proposed CCHP system can be applied using different operational strategies, which may be based on either thermal energy (heating/cooling) or electricity demand. However to allow complete utilization of the generated thermal energy, the system must be based on a thermal energy-led operation. Therefore if it is assumed that the system operates at a constant, full power capacity (which also allows maximum net electrical eff ciency), all generated thermal energy is delivered to buildings. The system is grid-interconnected to allow import/export of electricity, while the buildings are equipped with vapor-compression heat pumps, which will be operating, when the thermal energy suppl ed from the CCHP system is insufficient. The heat pumps are operated with electricity generated from the CCHP system. In the case of a CCHP system it is important to choose the set of consumers (buildings) that will allow an efficient matching of the generated vectors of energy with demand. The reason is that for example in the case of heating mode, high heat is required during the evening and early morning hours in households (due to high occupancy and low ambient temperature), while high heat demand occurs during daytime working hours in weekdays.

The system is assumed to be more applicable in locations with hot climates, in order to make full utilization of its cooling capability. Such a climate exists in the island of Cyprus, which is chosen as the location of application of the proposed system. In this case the load profile is generated based on local meteorological data ta<en from Florides et al. (2003). It is assumed that the load profile consists of three average building types: (1) household, (2) office bui ding, (3) large commercial building (e.g. shopping mall). The methodology suggested by Parker (2003) allows prediction of the load profile based on ambient temperature conditions. This methodology is used to size the thermal loads, with three operational levels: low (50%), medium (75%) and high (100%). For the household, maximum heating and cooling loads are 6 and 10 kW, respectively. The office building is assumed to require 20 times more thermal energy, while the commercial building requires 400 times more. The average COP-value of the vapor-compression heat pumps is taken to be equal to 3.0. To illustrate the pattern of operation of the selected load profile the two most extreme weather cond tions are shown in Table 11 and Table 12. The tables show the operational

pattern for an average day in January and July, based on expected demand for the total heating and cooling loads of each building type. The load profile includes 4200 households, 20 office buildings and 2 commercial buildings. The results show that the proposed system is able to fully utilize the generated thermal energy, while excess demand in thermal energy is satisfied by the heat pumps with electricity generated from the CCHP system.

Table 11: Pattern of operation for the buildings included in the load profile in January (heating mode)

Time of day	House-holds	Office Build-ings	Com-mercial Buildings	Total heating load demand	Total heating load satisfied by heat pumps	Total power required to drive the heat pumps[a]	Total power left for other usage[b]
00:00	18.9	0.0	0.0	18.9	0.9	0.3	12.5
01:00	18.9	0.0	0.0	18.9	0.9	0.3	12.5
02:00	18.9	0.0	0.0	18.9	0.9	0.3	12.5
03:00	18.9	0.0	0.0	18.9	0.9	0.3	12.5
04:00	18.9	0.0	0.0	18.9	0.9	0.3	12.5
05:00	18.9	0.0	0.0	18.9	0.9	0.3	12.5
06:00	25.2	0.0	0.0	25.2	7.2	2.4	10.4
07:00	25.2	1.2	2.4	28.8	10.8	3.6	9.2
08:00	18.9	2.4	3.6	24.9	6.9	2.3	10.5
09:00	12.6	2.4	4.8	19.8	1.8	0.6	12.2
10:00	12.6	2.4	4.8	19.8	1.8	0.6	12.2
11:00	12.6	2.4	4.8	19.8	1.8	0.6	12.2
12:00	12.6	1.8	3.6	18.0	0.0	0.0	12.8
13:00	12.6	1.8	3.6	18.0	0.0	0.0	12.8
14:00	12.6	1.8	3.6	18.0	0.0	0.0	12.8
15:00	12.6	1.8	3.6	18.0	0.0	0.0	12.8
16:00	12.6	1.8	3.6	18.0	0.0	0.0	12.8
17:00	18.9	1.2	3.6	23.7	5.7	1.9	10.9
18:00	25.2	0.0	3.6	28.8	10.8	3.6	9.2
19:00	25.2	0.0	3.6	28.8	10.8	3.6	9.2

20:00	25.2	0.0	3.6	28.8	10.8	3.6	9.2
21:00	25.2	0.0	3.6	28.8	10.8	3.6	9.2
22:00	25.2	0.0	3.6	28.8	10.8	3.6	9.2
23:00	25.2	0.0	3.6	28.8	10.8	3.6	9.2

[a] Heat pumps are assumed to operate at an average COP-value of 3.0.

[b] Electrical load profile and/or import/export to grid.

Table 12: Pattern of operation for the buildings included in the load profile in July (cooling mode)

Time of day	House-holds	Office Build-ings	Com-mercial Build-ings	Total cooling load demand	Total cooling load satisfied by heat pumps	Total power required to drive the heat pumps[a]	Total power left for other usage[b]
00:00	21.0	0.0	0.0	21.0	0.0	0.0	12.8
01:00	21.0	0.0	0.0	21.0	0.0	0.0	12.8
02:00	21.0	0.0	0.0	21.0	0.0	0.0	12.8
03:00	21.0	0.0	0.0	21.0	0.0	0.0	12.8
04:00	21.0	0.0	0.0	21.0	0.0	0.0	12.8
05:00	21.0	0.0	0.0	21.0	0.0	0.0	12.8
06:00	21.0	0.0	0.0	21.0	0.0	0.0	12.8
07:00	21.0	2.0	4.0	27.0	6.0	2.0	10.8
08:00	21.0	2.0	4.0	27.0	6.0	2.0	10.8
09:00	21.0	2.0	4.0	27.0	6.0	2.0	10.8
10:00	21.0	3.0	6.0	30.0	9.0	3.0	9.8
11:00	31.5	3.0	6.0	40.5	19.5	6.5	6.3
12:00	31.5	4.0	8.0	43.5	22.5	7.5	5.3
13:00	31.5	4.0	8.0	43.5	22.5	7.5	5.3
14:00	31.5	4.0	8.0	43.5	22.5	7.5	5.3
15:00	31.5	3.0	8.0	42.5	21.5	7.2	5.6
16:00	31.5	3.0	8.0	42.5	21.5	7.2	5.6
17:00	31.5	3.0	8.0	42.5	21.5	7.2	5.6
18:00	42.0	2.0	6.0	50.0	29.0	9.7	3.1

19:00	42.0	0.0	6.0	48.0	27.0	9.0	3.8
20:00	21.0	0.0	6.0	27.0	6.0	2.0	10.8
21:00	21.0	0.0	6.0	27.0	6.0	2.0	10.8
22:00	21.0	0.0	4.0	25.0	4.0	1.3	11.5
23:00	21.0	0.0	4.0	25.0	4.0	1.3	11.5

[a] Heat pumps are assumed to operate at an average COP-value of 3.0.

[b] Electrical load profile and/or import/export to grid.

Exergy Analysis

Table 7 shows the results for chemical exergy, physical exergy and total exergy, for summer and winter operation. The simulation results for the parameters of the exergy analysis defined in Section 4 are tabulated in Table 13. Exergetic efficiency varies depending on season, and is 0.443 during winter operation and 0.395 during summer operation. Therefore the average value for second law efficiency is 0.419, which is comparatively lower than the first law efficiency, i.e. the PER-value (0.91). This is due to the fact that exergy analysis is based on the qualitative evaluation of energy (energy availability), as compared to a plain quantitative evaluation. Therefore the exergy rate by recovery of waste heat to generate district heating and district cooling, which is low quality energy, results in a low exergy rate, and therefore low exergetic efficiency. Solving the equations given in Table 5, results in a total exergy destruction rate of 21.2 MW, out of 37.2 MW of exergy rate input to the system, while exergy loss is 0.7 MW. The equivalent entropy generation ($\dot{S}_{gen}$) is 71.8 kW/K.

Table 13: Results of exergy analysis for the proposed CCHP system

Variable description (unit)		Value
$\dot{X}_{el}$	Exergy rate of net electrical output	12.8 MW

$\dot{X}_h$	Exergy rate of district heating (winter mode)	3.3 MW
$\dot{X}_c$	Exergy rate of district cooling (summer mode)	1.7 MW
$\dot{X}_{fuel}$	Exergy rate of fuel	36.5 MW
η_2	Second law efficiency	0.419
$\dot{X}_{total}$	Total exergy rate	37.2 MW
$\dot{X}_{des}$	Total exergy destruction rate	21.2 MW
$\dot{X}_{loss}$	Loss in exergy rate	0.7 MW
$\dot{S}_{gen}$	Total rate of entropy generation	71.8 kW/K

By use of exergy destruction ratios, the exergy destroyed throughout the various processes in the system, an evaluation of the most significant contributors of exergy destruction can be made. These ratios are shown graphically in Fig. 5 and F g. 6, for summer and winter operation, respectively. Both figures suggest that the greatest portion of exergy destruction occurs primari y in the combustor. Secondary, yet significant, sources of exergy destruction are the heat exchangers recovering heat for district heating (HEx3), or district cooling (HEx2), due to the high temperature gradient in the flue gas side of each heat exchanger. The exergetic efficiency of the equivalent conventional system is only 0.336. Therefore cogeneration increases the exergetic efficiency of the system by almost 25%. The exergetic efficiency is however significantly lower than the PER, due to the fact that first law analysis does not distinguish between low and high quality energy. In other words, the generated energy is a sum of thermal energy (heat/cool) and electricity, which can only be assessed qualitatively by a second law analysis (Dinçer and Rosen, 2007). Therefore exergy analysis can provide more meaningful results, especially in the case of a mixture of different energy forms (i.e. electricity, heating, cooling). Specifically in the case of the proposed system, we can conclude that the system must be redesigned, with emphasis on the cooling plant, by consideration of different types of absorption chillers and/ or optimization of the current operational range of temperature. The gas turbine and the air compressor have low ratios of exergy destruction. Finally the exergy loss ratio is rather low, which means that the system is able to operate within minimum losses to the surroundings.

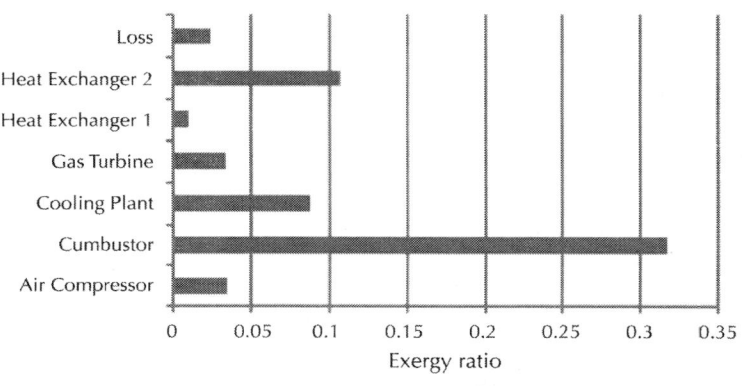

Figure 5: Exergy destruction and exergy loss ratios (summer operation).

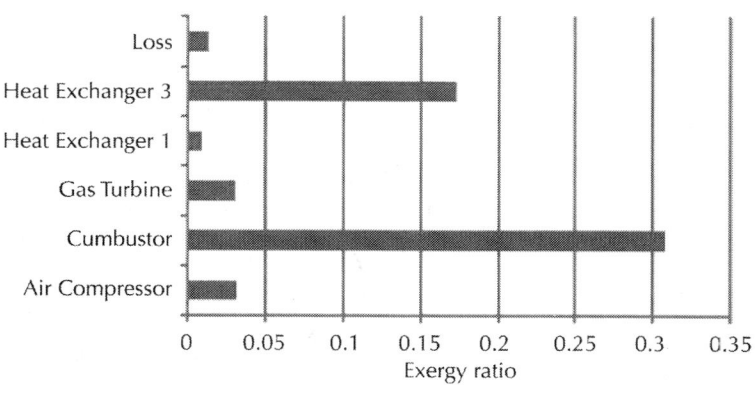

Figure 6: Exergy destruction and exergy loss ratios (winter operation).

CONCLUSIONS

A proposed small-scale, decentralized CCHP system is considered as alternative to conventional large-scale centralized, electricity-only generating power plants. The system is designed, modeled and analyzed in terms of thermodynamics, including cost analysis and exergy analysis. The results of the system model simulation show the potential of the proposed system in terms of efficiency. The system performs efficiently, with a PER of 0.91 and an exergetic efficiency of 0.419, and therefore the proposed system may become a significant candidate in the energy market, since fuel consumption and CO_2 emissions can be reduced. The considered fuel is LNG, and with the proposed design approach of its regasification, the cooling energy can be used to cool the feed air before compression, which results to a net electrical efficiency of almost 2%, which is an attractive system modification for areas with prolonged summer-like weather conditions. Overall, when the proposed system is compared to an equivalent conventional system, results show significant improvement in CO_2 emissions reduction (38.9%), and primary energy savings (41.8%). In terms of cost, the system results in a PP of approximately 4 years, with a total initial cost of 17.1 million .

The exergy analysis shows that there is a potential for improvement of the exergetic efficiency through design modifications. Exergy destruction is higher in the combustor, the cooling plant and the heat exchangers responsible for recovering waste heat (i.e. HEx2 and HEx3). In the case of the cooling plant and the heat exchangers, the irreversibility is due to the conversion of high quality waste heat to low quality useful heating (or cooling). Obviously this is an unavoidable consequence due to the purpose of the proposed system, although exergetic efficiency could improve with the replacement of the double-effect absorption chiller with a triple-effect one. Also, theoretically, net electrical efficiency (and also exergetic efficiency) can be improved with the introduction of a combined cycle configuration, which is however a more complex and costly, and thereby an unfavorable option for distributed generation applications. The only possible and feasible modification for the current system design could be the optimization of the combustion process, by preheating the reactants or minimizing the use of excess air, although this could result in an impractical (or inflexible) system configuration. Also it should be noted that a clear advantage of the proposed system is the low exergy rate losses to the surroundings. Overall the proposed system design achieves a high level of utilization of the supplied chemical energy of the fuel, while maintaining a compact design and a practical operational scheme.

REFERENCES

1.   Arsalis, A., 2012. Modeling and simulation of a 100 kWe HT-PEMFC subsystem integrated with an absorption chiller subsystem. Int. J. Hydrogen Energy 37, 13484e13490.

2.   Beith, R., 2011. Small and Micro Combined Heat and Power (CHP) Systems. Woodhead Publishing Limited, Cambridge, UK.

3.   Bejan, A., Tsatsaronis, G., Moran, M.J., 1996. Thermal Design and Optimization. Wiley.

4.   Deng, J., Wang, R.Z., Han, G.Y., 2011. A review of thermally activated cooling technologies for combined cooling, heating and power systems. Prog. Energy Combust. Sci. 37, 172e203.

5.   Dinçer, I., Rosen, M.A., 2007. Exergy: Energy, Environment and Sustainable Development. Elsevier Science Ltd, Oxford, UK.

6.   Dinçer, I., Zamfirescu, C., 2011. Sustainable Energy Systems and Applications. Springer, New York, NY.

7.   ESN-SECA, 2013. ESN e Way Forward SECA Report [WWW Document]. http://www. shortsea.info/openatrium-6.x-1.4/sites/default/files/esn-seca-report-2013_0.pdf   (accessed 31.07.14.).

8. Federal Energy Regulatory Commission, 2013. Other Markets: LNG - Imports, Sendouts, & World Prices [WWW Document]. http://www.ferc.gov/marketoversight/othr-mkts/lng.asp (accessed 05.11.13.).

9. Florides, G., Kalogirou, S., Theophilou, K., Evangelou, E., 2003. Analysis of the typical meteorological year (TMY) of Cyprus and house load simulation. In: Eighth International IBPSA Conference, Eindhoven, Netherlands, pp. 339e346.

10. GE, 2006. GE10-2 Gas Turbine [WWW Document]. http://www.ge-energy.com/ products_and_services/products/gas_turbines_small_heavy_duty/ge102_gas_ turbine.jsp (accessed 10.07.13.).

11. Gupta, S.B., 2012. Natural Gas: Extraction to End Use. InTech, Rijeka, Croatia.

12. Herold, K.E., Radermacher, R., Klein, S.A., 1996. Absorption Chillers and Heat Pumps. CRC Press, Boca Raton, FL.

13. Kakaç, S., Liu, H., 2002. Heat Exchangers: Selection, Rating, and Thermal Design, second ed. CRC Press, Boca Raton, FL.

14. Klein, S., Nellis, G., 2012. Thermodynamics. Cambridge University Press, New York, NY.

15. Kong, X.Q., Wang, R.Z., Huang, X.H., 2004. Energy efficiency and economic feasibility of CCHP driven by stirling engine. Energy Convers. Manage. 45, 1433e1442.

16. Linde, A.G., 2011. Baseload LNG Production in Stavanger. Linde Engineering, Pullach, Germany.

17. Liu, M., Lior, N., Zhang, N., Han, W., 2009. Thermoeconomic analysis of a novel zeroCO2-emission high-efficiency power cycle using LNG coldness. Energy Convers. Manage. 50, 2768e2781.

18. Mone, C.D., Chau, D.S., Phelan, P.E., 2001. Economic feasibility of combined heat and power and absorption refrigeration with commercially available gas turbines. Energy Convers. Manage. 42, 1559e1573.

19. Morosuk, T., Tsatsaronis, G., 2011. Comparative evaluation of LNG e based cogeneration systems using advanced exergetic analysis. Energy 36, 3771e3778.

20. Parker, D.S., 2003. Research highlights from a large scale residential monitoring study in a hot climate. Energy Build. 35, 863e876.

21. Popli, S., Rodgers, P., Eveloy, V., 2012. Trigeneration scheme for energy efficiency enhancement in a natural gas processing plant through turbine exhaust gas waste heat utilization. Appl. Energy 93, 624e636.

22. Shi, X., Agnew, B., Che, D., Gao, J., 2010. Performance enhancement of conventional combined cycle power plant by inlet air cooling, inter-cooling and LNG cold energy utilization. Appl. Therm. Eng. 30, 2003e2010.

23. Siemens, 2013. SGT-400 Industrial Gas Turbine [WWW Document]. http://www. energy.siemens.com/hq/en/fossil-power-generation/gas-turbines/sgt-400.htm   (accessed 10.09.13.).

24. Wang, J.-J., Jing, Y.-Y., Zhang, C.-F., Zhai, Z. (John), 2011. Performance comparison of combined cooling heating and power system in different operation modes. Appl. Energy 88, 4621e4631.

25. Wang, J., Yan, Z., Wang, M., Dai, Y., 2013. Thermodynamic analysis and optimization of an ammonia-water power system with LNG (liquefied natural gas) as its heat sink. Energy 50, 513e522.

A Liquefied Energy Chain for Transport and Utilization of Natural Gas for Power Production with CO_2 Capture and Storage – part 4: Sensitivity Analysis of Transport Pressures and Benchmarking with Conventional Technology for Gas Transport

Audun Aspelund and Truls Gundersen

The Norwegian University of Science and Technology, Department of Energy and Process Engineering, NO-7491 Trondheim, Norway

ABSTRACT

A novel energy and cost effective transport chain for stranded natural gas utilized for power production with CO_2 capture and storage is developed. It includes an offshore section, a combined gas carrier and an integrated receiving terminal. In the offshore section, natural gas (NG) is liquefied to LNG by liquid carbon dioxide (LCO_2) and liquid inert nitrogen (LIN), which are used as cold carriers. In the onshore process, the cryogenic exergy in the LNG is utilized to cool and liquefy the cold carriers, LCO_2 and LIN. The transport pressures for LNG, LIN and LCO_2 will influence the thermodynamic efficiency as well as the ship utilization; hence sensitivity analyses are performed, showing that the ship utilization for the payload will vary between 58% and 80%, and the transport chain exergy efficiency between 48% and 52%. A thermodynamically optimized process requires 319 kWh/tonne LNG. The NG lost due to power generation needed to operate the LEC processes is roughly one third of the requirement in a conventional transport chain for stranded NG gas with CO_2 capture and sequestration (CCS).

INTRODUCTION

The liquefied energy chain (LEC) is a novel energy and cost effective transport chain for stranded natural gas utilized for power production with CO_2 capture and storage, which includes an offshore section, a combined gas carrier, and an integrated receiving terminal, see Fig. 1. In the offshore section, natural gas (NG) is liquefied to LNG by liquid carbon dioxide (LCO_2) and liquid inert nitrogen (LIN), which are used as cold carriers. The nitrogen is emitted to the atmosphere at ambient conditions. The CO_2 at high pressure is transferred to an offshore oilfield for enhanced oil recovery (EOR). LNG is transported to the receiving terminal in the combined carrier.

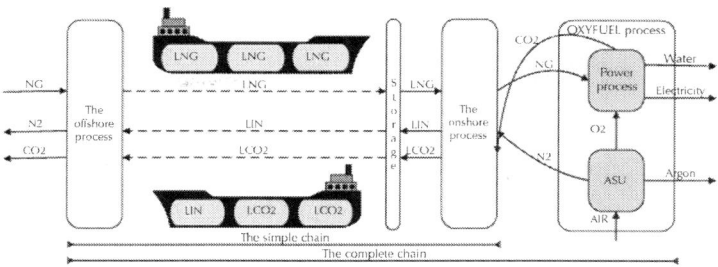

Figure 1: The liquefied energy chain.

At the receiving terminal, the cryogenic exergy in LNG is recovered by liquefaction of CO_2 and nitrogen. The onshore process is connected to an air separation unit (ASU) that produces nitrogen for the offshore process and oxygen for an oxyfuel power plant, where NG is converted to electricity, CO_2 and water. The water is removed by condensation from the CO_2 which is compressed to a pressure above the triple point (TP) and liquefied by vaporization of the remaining LNG. The LCO_2 and LIN are transported offshore in a combined gas carrier. Transporting CO_2 and LNG in the same ship results in an enhanced ship utilization. This paper is the last in a series of four papers that describe the liquefied energy chain. The first paper describes the concept and summarizes the results from the remaining three [1]. The second paper addresses the offshore and onshore processes [2]. The third paper describes the combined carrier [3]. This paper contains a general description of the thermo-mechanical exergy of LNG, LIN and LCO_2, sensitivity analyses of the ship utilization and exergy efficiencies as a function of transport pressures, as well as a benchmarking against conventional technologies for gas transport.

METHODOLOGY AND EXERGY ANALYSIS

The calculations in the first section of this paper, a general description of the effect of transport pressure, are based on pure components, and the data are collected from NIST [4]. In the rest of the paper, the processes are simulated in HYSYS [5], using the SRK equation of state. SRK is used as it is a well-known EOS for calculation of light hydrocarbons and pure components. We have tested other EOS, and the deviation is marginal. However, if a mixture of water, hydrocarbons and CO_2 is to be separated, the deviation will be very large. The NG composition is lean and consists of 1% nitrogen, 92% methane, 5% ethane, 1.8% propane and 0.2% butanes. Impurities such as mercury, water and CO_2 are assumed not present in the feed gas. The CO_2 and nitrogen streams are treated as pure gases. All processes are based on state-of-the-art equipment with standard industrial efficiencies. Additional information about of the simulation tools, equipment and ambient data can be found in the detailed description of the processes [2].

Neglecting the contributions from kinetic and potential energy and having no reactions in or mixing of the fluids, the specific change of exergy in a stream j through a unit and the exergy (rational) efficiency can be expressed as [6]

$$\Delta \varepsilon_j^{(tm)} = (h_{Outlet} - h_{Inlet})_j - T_0(s_{Outlet} - s_{Inlet})_j = (\varepsilon_{Outlet} - \varepsilon_{Inlet})_j \tag{1}$$

$$\psi = \frac{\sum_{out} m_{out}\varepsilon_{out} + E_{out}}{\sum_{in} m_{in}\varepsilon_{in} + E_{In}} \tag{2}$$

In Eqs. (1) and (2), is the specific exergy, h is the specific enthalpy, s is the specific entropy and m is the mass flow of a stream. E is the exergy supplied or removed from the system. T_0 is the ambient temperature. The exergy (rational) efficiency, , describes the fraction of the input exergy (available energy) in a process that is converted to useful exergy. The exergy conversion efficiency is defined as the exergy efficiency taking into account only the exergy components that change throughout the process. In a steady-state NG process with no mixing, separation or chemical reactions, the conversion exergy efficiency will neglect the chemical exergy and focus on the thermo-mechanical components. Hence, the conversion exergy will be lower, but more representative, than the total exergy efficiency which includes the chemical components in both the inlet and the outlet streams (see Fig. 2).

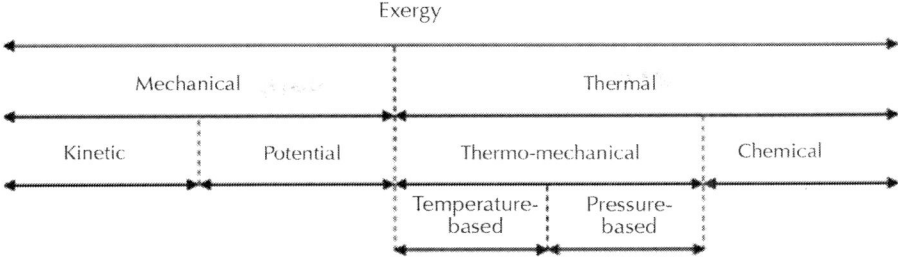

Figure 2: Components of exergy.

In thermodynamics, the exergy can be divided into physical and chemical exergy. The physical exergy can be further divided into mechanical exergy, which is normally neglected in process calculations, and thermo-mechanical exergy. The thermo-mechanical exergy component can be split into temperature and pressure based exergy. The temperature based exergy $^{(T)}$ is defined as the maximum obtainable work when a system, such as a stream is brought from its current temperature (T) to ambient temperature (T_0) at constant pressure (P). Pressure based exergy $^{(P)}$ is the maximum obtainable work when the stream is brought from its current pressure (P) to ambient pressure (P_0) at ambient temperature (T_0), hence [6]

$$\varepsilon^{(T)} = h(T,P) - h(T_0,P) - T_0(s(T,P) - s(T_0,P)) \tag{3}$$

$$\varepsilon^{(P)} = h(T_0,P) - h(T_0,P_0) - T_0(s(T_0,P) - s(T_0,P_0)) \tag{4}$$

If ideal gas with constant heat capacity C_p is assumed, the expressions for temperature (Eq. (5)) and pressure based exergy (Eq. (6)) become [6]

$$\varepsilon^{(T)} = C_p\left[T - T_0\left(1 + \ln\frac{T}{T_0}\right)\right] \tag{5}$$

$$\varepsilon^{(P)} = T_0 R \ln\frac{P}{P_0} = \frac{k-1}{k}C_p T_0 \ln\frac{P}{P_0} \tag{6}$$

The total specific chemical exergy for a fuel consists of the net change of Gibbs free energy of formation between the products and the reactants, the ideal work generated by expansion of the products and inert gases to their partial pressures in the atmosphere as well as a negative contribution for the ideal work needed to compress the reactants and inert gases from their original partial pressures in the fuel mixture or air to atmospheric conditions. Due to this definition of chemical exergy, it is more difficult to divide it into mixture-based exergy and chemical potential. For a pure ideal atmospheric gas, the molar mixture-based exergy can be expressed by Eq. (7). For a mixture of gases the molar mixture-based exergy can be expressed as Eq. (8). The chemical exergy of one component of the fuel is defined as Eq. (9), where $-\Delta_f G_i^\circ$ is the net change in Gibbs free energy of formation at standard conditions (1 atm and 25 °C) and ν_j is the stoichiometric coefficient of each species in the reaction with negative values for the reactions and positive values for the products. By combining Eqs (7), (8) and (9) the total chemical exergy for a mixture of gases can be found. A more detailed description can be found in [6] and [7].

$$\varepsilon_{i,0}^{(ch)} = -RT_0 \ln X_i^e \tag{7}$$

$$\varepsilon_{mix}^{(ch)} = \sum_i X_i \varepsilon_i^{(ch)} + RT_0 \sum_i X_i \ln X_i \tag{8}$$

$$\varepsilon_i^{(ch)} = -\Delta_f G_i^\circ + \sum_{j \neq i} \nu_j \varepsilon_j^{(ch)} \tag{9}$$

In order to calculate the exergy content of a fuel, the ambient surroundings and the composition of the atmosphere must be known for reference. In the calculations, an ambient air temperature of 25 °C, an atmospheric pressure of 1 bar ($\approx$1 atm), a humidity of 70%, and the atmospheric composition given in the US Standard of atmosphere from 2004 is used [4]. Table 1 shows the fuel composition, the atmospheric composition and the chemical exergy for each component. Note that the last term in Eq. (9) will be a negative term for the fuel, as each component has a lower partial pressure than 1 bar. As a curiosity, the expected increase of

CO_2 in the atmosphere from 375 ppm to 550 ppm will decrease the chemical exergy of the current fuel with 0.12%. However, the temperature and humidity are by far more important. A discussion of chemical exergy, with sensitivity analyses for atmospheric gaseous fuels to variations in ambient conditions, was recently presented by Ertesvåg [7].

Table 1: Chemical exergy of natural gas

Component	Fuel (mol%)	Content in atmosphere (mol%)	$\Delta\varepsilon^{(ch)}$ As pure (kJ/mol)	$\Delta\varepsilon^{(ch)}$ fuel (kJ/mol)
Nitrogen, N_2	1	77.1185	0.7	−0.1
Methane, CH_4	92	–	831.5	764.9
Ethane, C_2H_6	5	–	1495.5	74.4
Propane, C_3H_8	1.8	–	2151.1	38.5
n-Butane, C_4H_{10}	0.2	–	2805.0	2.8
Water, H_2O	0	2.1619	9.5	2.7
Carbon dioxide, CO_2	0	0.0370	3.9	–
Oxygen, O_2	0	20.6826	274.9	–
Total	100	100.0000	–	883.2
Fuel total exergy (kWh/tonne)	–	–	–	14 057

The common basis for calculating efficiencies of power processes is to use the lower heating value (LHV). However, the LHV is not the same as the chemical exergy and should therefore not be used in exergy calculations. With the current gas specification, the LHV is 48.8 MJ/kg, whereas the chemical exergy is 50.6 MJ/kg, meaning that the chemical exergy is 3.7% higher than the LHV, which is within the normal difference of 3–5%. The total exergy efficiencies given in this paper will therefore be roughly 2% lower than stated in other papers using LHV as the basis. The efficiencies in the power processes are reduced accordingly, e.g. a power process with a LHV efficiency of 50% will have an exergy efficiency of 48%.

Finally, in contradiction to most papers, the thermo-mechanical exergy of the fluids entering and leaving the system is included in the exergy calculations. For example the total exergy of a NG stream at ambient temperature and a pressure of 70 bar consists of 99% chemical exergy and 1% thermo-mechanical exergy.

THE ENERAL INFLUENCE OF TRANSPORT PRESSURE

Due to the TP of CO_2 at 5.2 bar, CO_2 cannot be a liquid at lower pressures. This has two implications: first, the combined gas carrier must be semi-pressurized. Second, the minimum temperature of liquid CO_2 cannot be lower than $-$ 56.6 °C. Semi-pressurized gas carriers enable LNG to be transported at a higher pressure and temperature; hence the required cooling is less than for ordinary LNG transport. Semi-pressurized vessels also give a degree of freedom in the transport pressure for LIN. This can be exploited to maximize the total chain efficiency and the ship utilization.

In the LEC concept, two different density types are important. The mass density determines how much liquid that can be transported in the vessel. The exergy density determines how much thermo-mechanical exergy that can be transported for a given volume. As can be seen from Fig. 3, the densities of LNG, LCO_2 and LIN all decrease with increasing pressure; hence, from a ship utilization point of view, the fluids should be transported at low pressures to obtain a higher volumetric efficiency. Also note that the mass density for LCO_2 is three times the density of LNG.

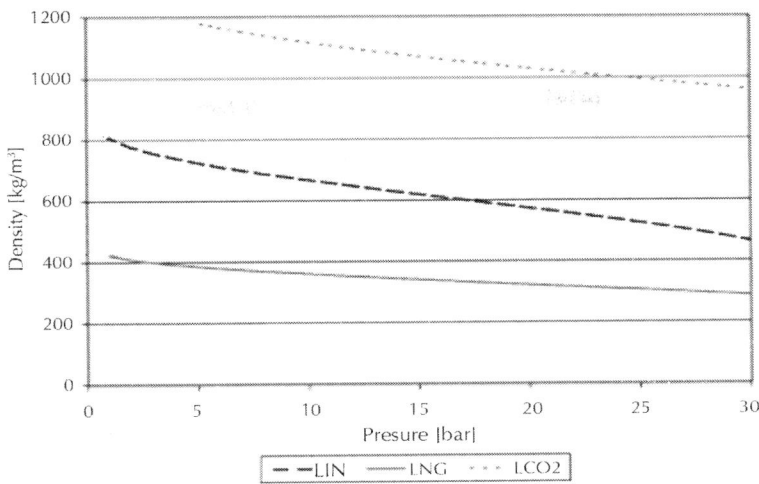

Figure 3: Mass densities for LNG, LIN and LCO$_2$ at the bubble point.

Since LCO$_2$ and LIN are used as cold carriers for the LNG production, the thermo-mechanical exergy change in the cold carriers must be larger than, or equal to, the exergy change in the NG stream. The thermo-mechanical exergy consists of pressure based and temperature based exergy. Fig. 4 shows the pressure based, temperature based and total thermo-mechanical exergy for LNG, LIN and LCO$_2$ in the range from 1 to 60 bar at bubble point conditions. The pressure exergy for the three fluids is more or less equal and increases rapidly from 1 bar and up to 10 bar, where the curves gradually flatten out. The temperature exergy is, as expected, highest for the most volatile gas, LIN, at low pressures. Note that the temperature based exergy for LNG at 1 bar equals that of LIN at 4 bar. The pressure exergy equals the temperature based exergy in both LIN and LNG at about 20 bar. Note that the total exergy for CO$_2$ hardly changes with pressure. The total exergy of CO$_2$ in dense phase at an injection pressure of 150 bar is only 1% higher than the exergy at a transport pressure at 5.5 bar. Therefore, in the CO$_2$ stream, the exergy is shifted from temperature based and pressure based exergy to pressure based exergy throughout the process, whereas the thermo-mechanical exergy in the inlet and outlet streams is about the same. The NG enters the process at about 70 bar, and will therefore require less exergy to be liquefied; however, even at high transport pressures, there is still a significant deficit of exergy, which has to be provided by the LIN.

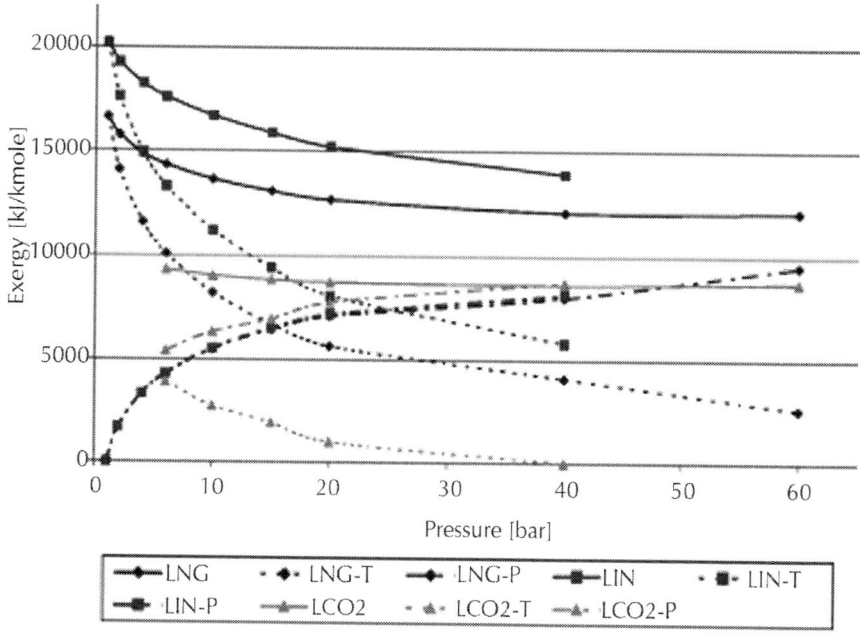

Figure 4: Total thermo-mechanical, pressure based (P) and temperature based (T) exergy for LNG, LIN and LCO_2 at bubble point pressure.

Fig. 5 shows the exergy density as a function of saturation pressure for LNG, LIN and LCO_2. LIN has the highest exergy density, with a steep increase from 5 bar to 1 bar. The exergy density of LNG is lower than for LIN and also increases significantly from 5 to 1 bar. Due to the high mass density, the exergy density of LCO_2 is relatively high, although the temperature is significantly higher.

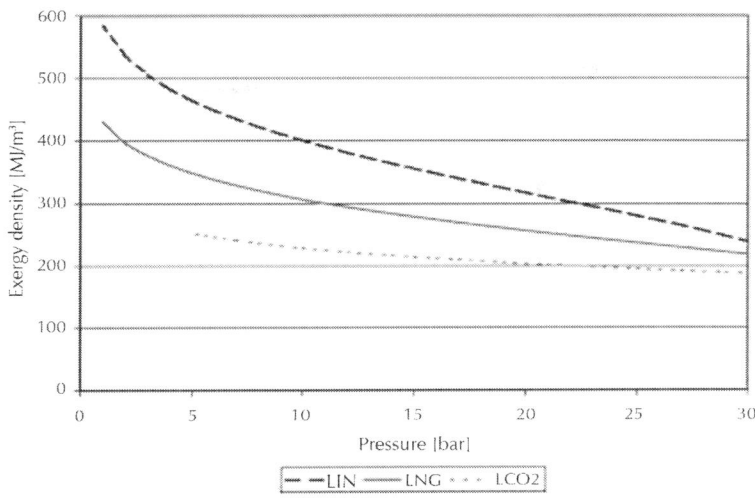

Figure 5: Exergy densities for LNG, LIN and LCO$_2$ at the bubble point.

From Fig. 5 it seems reasonable that a gas carrier filled about half with LCO$_2$ and the rest with LIN would be capable of liquefying a full shipload of LNG; however, this does not give the complete picture. On the positive side, the NG has an inlet pressure of 70 bar which provides exergy to the process; hence the NG will require less exergy to be liquefied. On the negative side, the CO$_2$ needs to be injected at 150 bar, which has approximately the same exergy as the liquid CO$_2$; hence, although cooling duty is provided, the net exergy contribution (outlet minus inlet) to the liquefaction of NG is rather small. Nitrogen, on the other hand, is emitted to the atmosphere at ambient conditions; therefore all exergy in the nitrogen can be utilized in the process. Finally there will be irreversibilities in the process, so the net inlet exergy in must be larger than the net outlet exergy. Fig. 6 shows the net exergy provided to the offshore process from LNG, LIN and LCO$_2$ as a function of bubble point pressure. Note that, compared to Fig. 5, the net exergy density is unchanged for the nitrogen, close to zero for the CO$_2$ and significantly reduced for the LNG.

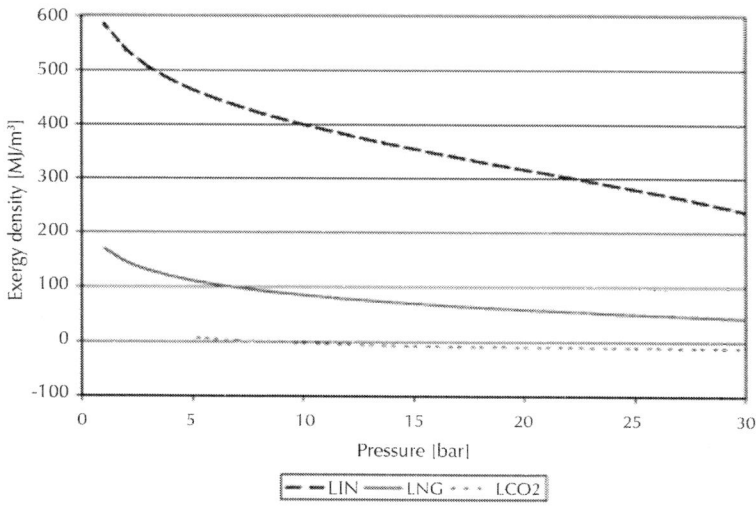

Figure 6: Net exergy provided to the process from LNG, LIN and LCO$_2$.

Based on the assumption that the offshore process is to be self-supported with exergy, the ship utilization can be calculated. The ship utilization factor (SUF) is a concept used to quantify the efficiency of the cargo containment system. The SUF shows to what degree the cargo space is used to transport moneymaking cargo, in shipping known as the "payload". In the LEC project, the payload is LCO$_2$ and LNG. The LIN is transported only to make the production of LNG possible, thus it is not generating income in itself. The conventional way of defining ship utilization is to divide the volume of transported cargo per year on the maximum yearly transport capacity for a given ship. Since the LEC ship carries load both ways it is more effective than a typical liquefied gas ship, and could obtain a conventional SUF higher than 100%. Therefore, contrary to standard SUF calculations, the SUF is defined by adding the volume of outbound and inbound transported payload and dividing it on the total cargo volume times two. It is, of course favourable to have as high ship utilization as possible, as this will reduce the void space in any direction. The ship utilization as a function LNG and LIN transport pressure is shown in Fig. 7. The LCO$_2$ is transported at 5.5 bar and the offshore process efficiency is 87% [2]. It is assumed that 85% of the carbon that is transported onshore in the NG is returned offshore as CO_2 for EOR. The calculations are based on the assumption of pure nitrogen and pure CO_2. Furthermore, the NG is represented as pure methane. As can be seen, the SUF will vary from 57% when LNG is transported at low pressure (1 bar), and LIN is transported at high pressure (15 bar), to 88% when LNG is transported at high pressure and LIN at low pressure. It should be emphasised that the processes described in [2] is designed for an LNG pressure of 1 bar and a LIN pressure of 6 bar for close to optimal utilization of the cryogenic exergy and has an efficiency of 87%. Since the process efficiency will change, and certainly decrease for many of the pressure combinations, Fig. 7 should only be used as a general guideline. A detailed description of the most favourable transport pressures and the effect on the process efficiencies and ship utilization factor can be found in the following section.

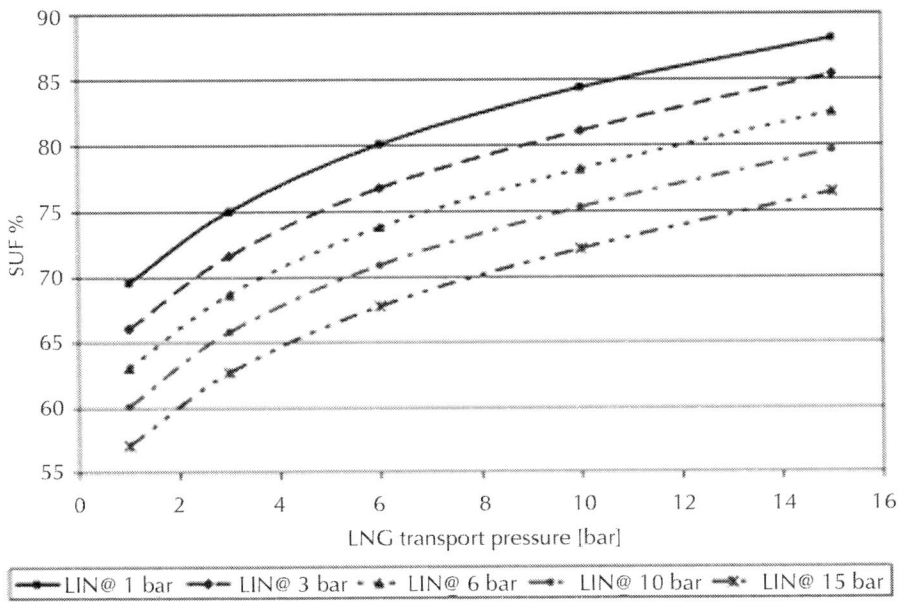

Figure 7: Ship utilization as a function of LNG and LIN transport pressure.

THE EFFECT OF TRANSPORT PRESSURE UPON SHIP UTILIZATION AND PROCESS EFFICIENCIES

As discussed in the previous section, selecting the correct transport pressure for LNG, LCO_2 and LIN is important for a high SUF. However, the transport pressure is also important for the exergy efficiencies for both the onshore, offshore and combined processes (the LEC). In this section, a detailed analysis of the exergy efficiencies for the offshore and onshore processes, as well as for the LEC, is evaluated together with the ship utilization for various transport pressures. The exergy efficiencies are calculated using the input data, processes, tools and thermodynamic methods described in [2]. Since the calculations here are done more rigorously using process calculations as well as a multi component NG mixture, the results will be slightly different from the simplified calculations in the previous section.

The calculations show that CO_2 should be transported at lowest possible pressure. There are two reasons for this: first, the density increases with decreasing pressure, giving better ship utilization. Second, there is a lack of cooling duty between −70 °C and −40 °C in the offshore process, whereas there is a surplus of cooling duty above −90 °C in the onshore process. However, as liquid CO_2 cannot exist below the TP (5.2 bar), the minimum pressure is assumed to be 5.5 bar. This is already lower than recommended for ship based transport of CO_2 [10].

The LNG and the LIN can be transported at any pressure between 1 bar and maximum vessel pressure. Increasing the pressure of LNG will decrease the need for cooling duty offshore. However, the LNG will then provide less cooling at a higher temperature at the receiving terminal. As CO_2 needs to be transported in semi-pressurized tanks, the LNG and LIN can be transported up to 6 bar without additional costs. Above this threshold there will be an increase in the building costs of the combined carrier.

In a thermodynamically optimized transport, that is a transport chain with high exergy efficiency, the selected transport pressure is 5.5 bar for CO_2 and 1 and 6 bar for LNG and nitrogen, respectively. In order to be self-supported with power at these pressures, 1 m^3 of produced LNG requires 0.61 m^3 LIN and 0.86 m^3 CO_2, giving a SUF of 63.4%. Hence a 12 250 m^3 ship will be able to transport 8330 m^3 LNG to shore and requires 7180 m^3 LCO_2 and 5070 m^3 LIN from shore to the field. Note that the required CO_2 is less than the carbon from the NG, so if all the CO_2 is captured and transported back to the reservoir, the ship utilization will actually decrease as 1 mol of the specified NG gives 1.076 mol of CO_2. By decreasing LIN transport pressure, more cooling duty can be transported offshore in the same volume, thereby increasing the ship utilization. However, there will be a larger negative temperature gap between the LNG to be vaporized and the nitrogen to be condensed at the market site. From an exergy point of view the bubble point temperature for LIN should be slightly below the bubble point of LNG, as this will give the smallest losses in the heat exchangers, both in the offshore process and for the simple transport chain.

Fig. 8 shows the SUF and the required amount of LCO_2 and LIN in the offshore process as a function of LNG pressure and maximum ship pressure. On a similar basis, Fig. 9 shows the exergy efficiencies for the offshore, onshore and simple transport chain, versus LNG and maximum ship pressure.

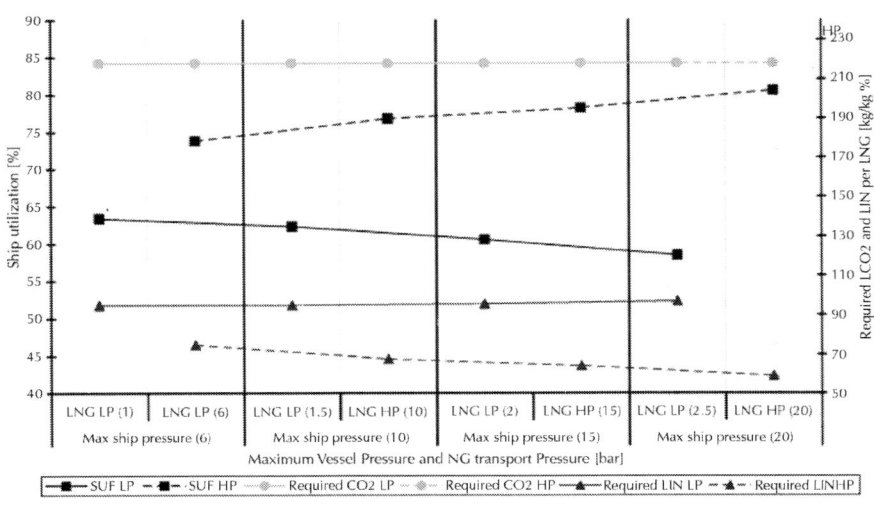

Figure 8: Ship utilization and required LCO_2 and LIN as a function of ship- and LNG pressure.

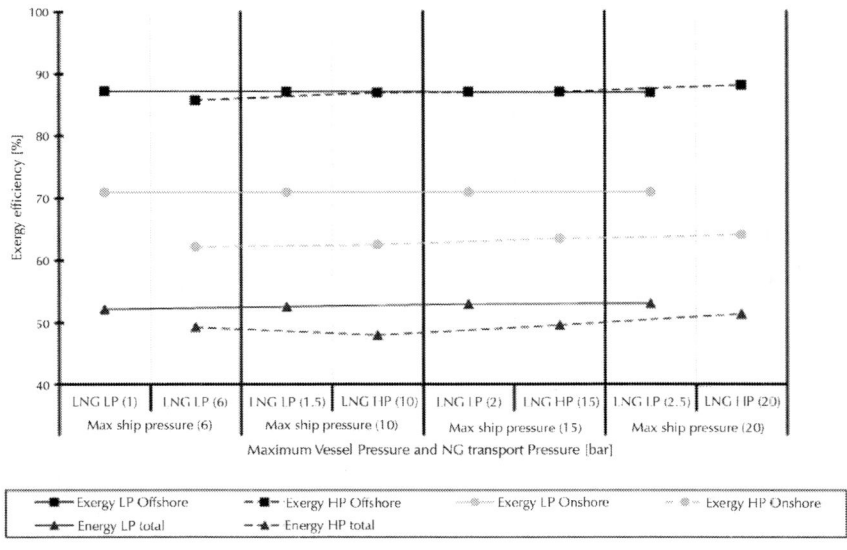

Figure 9: Exergy efficiencies for the offshore process, the onshore process and the simple transport chain as a function of ship- and LNG pressure.

The LCO$_2$ and LIN are transported at 5.5 bar and maximum vessel pressure, respectively. The LNG is transported at a low pressure (LP) which is the most energy-efficient pressure. Alternatively it is transported at maximum ship pressure (HP) to minimize the use of LIN. As an example, a graph for the required LIN for transport of HP LNG at increasing maximum vessel pressure can be seen at the bottom ofFig. 8. The required amount of CO$_2$ is constant at 2.21 kg/kg LNG. It is possible to reduce the amount of CO$_2$ at the cost of an increase in the required LIN and some simple process modifications, however, in this work the rate of CO$_2$ is kept constant in the simulations. Note that the amount of CO$_2$ corresponds to roughly 85% of the carbon in the LNG.

If the LNG is transported at LP, the nitrogen requirements will be almost constant. The density of nitrogen will decrease with increasing pressure; hence the ship utilization will decrease slightly. Due to the equal temperature difference between LNG and LIN, the exergy efficiency in the offshore process (87%) and the onshore process (71%) is also constant with increasing pressures. The simple transport chain efficiency (52%) is also constant for transport of LNG at low pressure. Note that an increase in LNG pressure from 1 to 2.5 bar leads to an increase in nitrogen pressure from 6 to 20 bar.

If both LNG and LIN are transported at the maximum ship pressure there is a substantial increase in ship utilization. At a maximum ship pressure of 6 bar, the ship utilization is 74%. It will continue to increase with increasing pressure, and is 80% at a vessel pressure of 20 bar. The reason for the increase in ship utilization is twofold. The nitrogen requirement decreases with increasing pressure. Also, the density of LNG decreases faster than the density of the LIN. As a result of the decrease in LNG density with increased pressure, the total amount of LNG transported increases less than the increase in ship utilization. At a transport pressure for LNG of 1 bar, 304 kg/m^3 can be transported, and the ship utilization is 63.4%. At 10 bar, 337 kg/m^3 is transported with a ship utilization of 77%, whereas the numbers for 15 bar are 350 kg/m^3 and 80%. Although there is an increase in irreversibilities in the heat exchangers due to larger temperature differences, the exergy efficiency in the offshore process will increase slightly with increasing maximum ship pressure. The reason is that the losses in the nitrogen path are smaller due to less nitrogen in circulation. However, the exergy efficiency of the onshore process drops by 9% points from 71% to 62%. This is due to an increased negative temperature gap between LIN and LNG, giving large irreversibilities since it is combined with an increase in recycled nitrogen flash gas. For the simple chain, there is a decrease in efficiency by approximately 4% points from 52% to 48% when LNG is transported at high pressure.

One point of particular interest is the connection between process simulations and the overall optimization. The optimal transport pressure is a function of the distance from field to marked, the CO_2 capture rate, the NG or electricity price, and the cost of the gas carriers. However, since the process efficiency also changes with transport pressure, this should have been taken into account already in the design phase. Two approaches are possible to overcome this problem. The efficiencies can be linearized as a function of the transport pressures and used in the overall optimization routine; alternatively the optimization routine can be connected to the HYSYS simulation so that the process is optimized along with the overall market determined variables. The main challenge in the latter approach is that convergence properties in the simulation may cause disturbances for the optimizer, thus gradient methods may fail to find the optimum.

BENCHMARKING WITH CONVENTIONAL GAS TRANSPORT PROCESSES

Fig. 10 describes a complete chain for electricity production with carbon capture and sequestration (CCS) based on NG. The main elements in this chain are; production and transport of NG, conversion of gas to electricity with CO_2 capture, and transport and injection of CO_2.

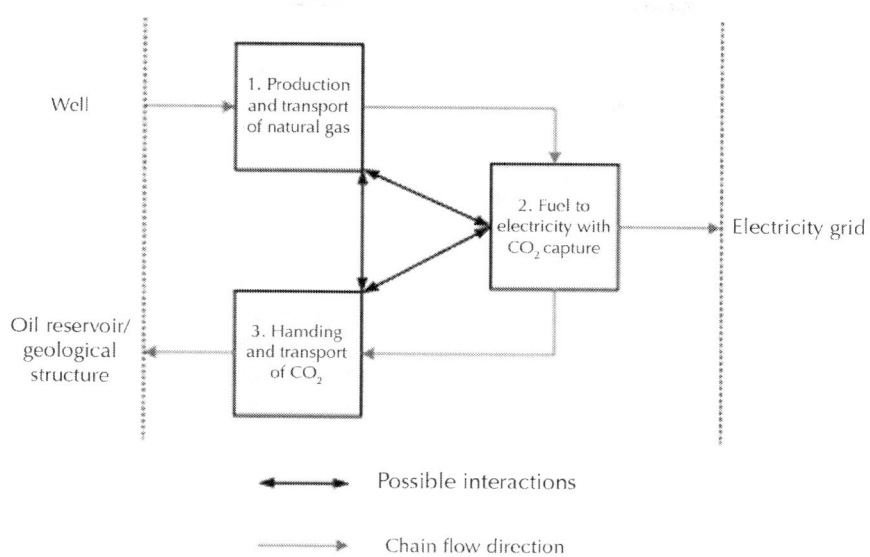

Figure 10: A complete chain from NG well to CO_2 storage.

Apart from the LEC (Fig. 1) where the elements in the complete chain are integrated, three different state-of-the-art transport chains are used in the benchmarking; a conventional chain for LNG transport withoutCCS (Chain A), a chain with ship based transport of LNG and LCO_2 (Chain B), and pipeline transport of both NG and CO_2, (Chain C). Fig. 11 shows the alternative energy chains A–C. The benchmarking includes the energy requirements for the processes in the simple chain that only includes transport. Also, the required loss of NG combusted in gas turbines to provide the required shaft work in a simple chain is presented, as power generation will have a different efficiency depending on where in the chain it is done. Finally, an exergy efficiency analysis of the full chain, including the power process for the LEC and the alternatives A–C, is presented in Table 3.

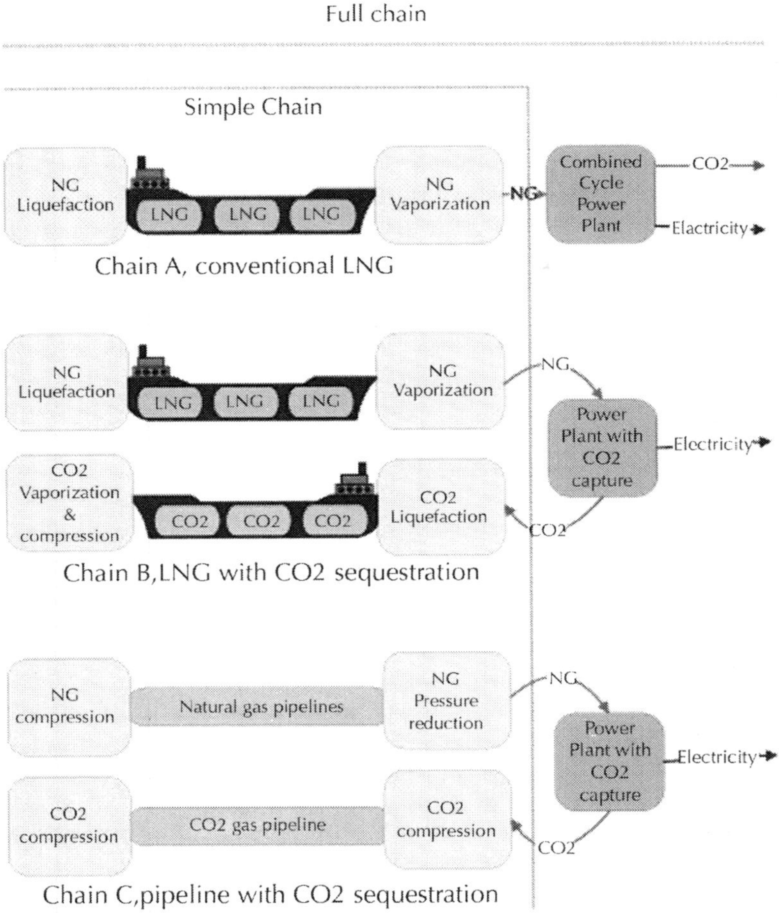

Figure 11: Alternative energy chains.

Table 3: Comparison of the efficiency of chain A–C and the complete energy chain

	The LEC (kWh/ tonne)	A, conv. LNG (kWh/tonne)	B, LNG and CCS (kWh/ tonne)	C, pipeline and CCS (kWh/ tonne)
NG@70 bar	156	156	156	156
NGLHV	14 057	14 057	14 057	14 057

Total exergy in	14 213	14 213	14 213	14 213
Losses offshore	0	1379	1441	176
Power prod.	6822	7059	6131	6738
Losses onshore	383	20	320	295
Net power out	6439	7039	5811	6443
CO_2@150 bar	153	0	153	153
Total exergy out	6592	7039	5964	6596
Exergy eff. (%)	46.4	49.5	42.0	46.4

Conversion of NG to Electricity

Offshore, the required power to the process can be produced by open cycle gas turbines with an exergy efficiency of 29% (LHV efficiency of ≈30%). The CO_2 emitted from the offshore power production cannot easily be captured. Onshore power is produced in a conventional combined cycle power plant with an exergy efficiency of 55% (LHV efficiency of ≈57%), or in a power plant with CO_2 capture. Three main concepts are described in the literature; pre-combustion, post-combustion and oxyfuel. A comparison of different capture technologies performed by Kvamsdal et al. shows that the average efficiency excluding transport of CO_2 is close to 50%, based on the LHV, which corresponds to an exergy efficiency of 48% [8]. Some capture processes do not capture all the CO_2, and the LEC will have a higher ship utilization if not all the CO_2 is returned. Therefore, the transport alternatives are calculated using a CO_2 capture rate of both 100% and 85%.

Gas conditioning of NG and CO_2 for Pipeline or Ship Transport

In this paper it is assumed that the NG and CO_2 gas is conditioned for transport. Pre-treatment of NG consists of water removal, mercury and CO_2 removal as well as a reduction in heavy hydrocarbons. Depending on the feed gas, the gas conditioning for pipeline and especially ship transport can be a very expensive process. As described in [2] the gas conditioning of NG can be integrated in the LEC and take advantage of the cold from the LCO_2. Furthermore, if NG is transported at a higher pressure, e.g. above 10 bar, the specifications on CO_2 and HHC may be relaxed.

The CO_2 to be transported will normally be saturated with water, which has to be removed down to 50 ppm. The CO_2 may also contain nitrogen, argon, small fractions of hydrogen or oxygen, which has to be removed within certain limits, to avoid corrosion or dry ice and hydrate formation. The gas conditioning of CO_2 for pipeline and ship transport from several different gas power plant processes with CO_2 capture is discussed in [14]. The impact of impurities in oxy-fuel combustion based processes is discussed in [15]. The gas conditioning of CO_2 can be included in the LEC concept with relatively small costs and modifications. However, when calculating these processes it is important to suitable thermodynamic models as water and hydrocarbons will change the vapor–liquid equilibrium in CO_2 rich mixtures significantly [16] and [17].

Ship Transport of NG and CO_2

In conventional ship transport of NG without CO_2 capture (Chain A), the NG is liquefied and stored on an floating production storage and offloading vessel (FPSO) before it is unloaded to an LNG carrier, which transports the LNG onshore to the receiving terminal. A description of the most important features for the selection of offshore LNG processes and a comparison of energy efficiencies with the propane precooled mixed refrigerant, (C3-PMR) can be found in [9]. The energy requirements range from roughly 300 kWh/tonne LNG for the base load C3-PMR process, to over 900 kWh/tonne LNG for the single closed nitrogen expander cycles that are considered more suitable for offshore production. The C3-PMR is developed by the world leading LNG process licenser, Air Products and Chemicals Incorporated, (APCI), and is the most widely used LNG process in the world. Another promising base load LNG process, the mixed fluid cascade (MFC) process developed by Statoil and Linde, is used in a base load LNG plant at Hammerfest, and is in the upper end when it comes to process efficiency with energy requirements between 250 and 300 kWh/tonne LNG depending on the ambient temperature and feed gas pressure [10]. However, both of these processes are originally designed for onshore base load LNG plants, hence there are some difficulties in transferring the technology to offshore production.

An offshore LNG process, named Niche-LNG, with a dual expander loop and a predicted energy requirement of 400 kWh/tonne, is developed by ABB [11]. This nitrogen expander process is especially designed for large-scale offshore LNG processes since it can use plate-fin heat exchangers and does not require flammable refrigerants. It should be noted that the efficiency is in the high end of offshore LNG processes and that the process is not commercialized. Due to the many similarities with the technology in the LEC chain, Niche-LNG is used in the benchmarking. At the receiving terminal, the LNG is pumped to storage tanks, vaporized and supplied to a combined cycle power plant. The regasification efficiency varies significantly. A modest 20 kWh/tonne LNG is used in the calculations.

Chain B consists of the same steps as chain A, however, the NG is supplied to a power plant with CO_2 capture. The CO_2 is liquefied, stored and loaded to a CO_2 ship, which transports the CO_2 offshore. An offshore unloading system including pumping, heating and transfer of CO_2 to the injection well is required, however, no storage is needed offshore as batch injection actually is favorable for the oil recovery [12]. The gas specifications for ship transport are stricter than for pipeline transport both for NG and CO_2. The CO_2 is conditioned to transport specifications during the liquefaction process. Ship transport of CO_2 requires 110 kWh/tonne LCO_2 for liquefaction and 6.5 kWh/tonne LCO_2 for offshore unloading. The data for ship transport of LCO_2 are collected from [13]. The data are based on similar assumptions as in the current paper. If 85% of the CO_2 is returned, the energy requirements are 243 and 14.3 kWh/tonne LNG, respectively, for liquefaction and offshore unloading, whereas if all CO_2 is captured and sequestrated the numbers are 286 and 16.9 kWh/tonne LNG.

Pipeline Transport of NG and CO_2

In state-of-the-art pipeline transport, as described in Chain C, the NG is conditioned to dew point specifications and compressed before it is transported onshore in a pipeline. The onshore process consists of a gas power plant with CO_2 capture. After capture, the CO_2 is conditioned to transport specifications, compressed and transported offshore in pipeline where it is injected in an oil reservoir for EOR. Using similar assumptions as in [2], the energy requirement for compression of NG from 70 to 200 bar for transport is 51 kWh/tonne NG. On a similar basis, the direct compression of CO_2 from 1 to 250 bar is 113 kWh/tonne CO_2, which corresponds to 295 kWh/tonne LNG if all CO_2 is re-injected and 250 kWh if 85% of the CO_2 is returned. These figures are confirmed by [14]. The total energy requirement for the pipeline transport is 346 and 287 kWh/tonne LNG for 100% and 85% return of the CO_2, respectively.

The Simple Transport Chain with Combined Ship Transport of LNG and CO_2

In the simple transport chain, NG at 70 bar is processed and transported from the field site to the market site where it is delivered at 25 bar. CO_2 at atmospheric pressure is processed and transported from the market site to the field site where it is unloaded at 150 bar and at a temperature higher than 15 °C. LIN is used as a cold carrier. In the thermodynamically optimized process, only 85% of the carbon is required in the offshore process. Battery limits and exergy efficiency for the simple transport chain are found in Fig. 12. Nitrogen enters and leaves the system boundaries at ambient temperature and atmospheric pressure, and is therefore not included in the exergy calculations. The exergy efficiency for the simple transport chain is 52%. The offshore process is self-supported with power; however, the required energy in the onshore process is 319 kWh/tonne LNG. Production of nitrogen in an ASU will require 47 kWh/tonne nitrogen, which corresponds to 45 kWh/tonne LNG and will decrease the exergy efficiency to 48%. If all the CO_2 is returned for sequestration, the required energy including the ASU is 383 kWh/tonne LNG.

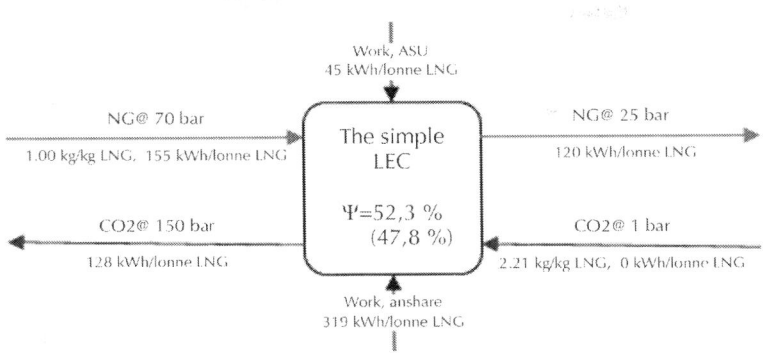

Figure 12: Battery limits and exergy efficiency for the simplified liquid energy chain.

A Comparison of the Simple LEC and Conventional Pipeline and Ship Transport of NG and Co$_2$

The energy requirements and the required gas used for power generation for the processes for the simple energy chain are shown in Table 2. A conventional ship based LNG chain with (Chain A) and without (Chain B) return of CO_2 and a pipeline chain (Chain C) are also shown for comparison. The energy requirements are given for 85% and 100% capture and sequestration fraction of CO_2.

Table 2: The simple liquefied energy chain vs. conventional chains

Returned CO$_2$		Energy req. (kWh/tonne)		Power cycle eff. (%)	Loss of NG (%)	
		100%	85%		100%	85%

The simple LEC	Offshore	0	0	29	0	0
	Onshore	338	319	48	5.0	4.7
	ASU	45	45	48	0.7	0.7
	Total	383	364		5.7	5.4
Chain A LNG without CCS	LNG prod.	400	400	29	9.8	9.8
	LNG receiv.	20	20	55	0.3	0.4
	Total	420	420		10.1	10.1
Chain B LNG with CCS	LNG prod.	400	400	29	9.8	9.8
	LNG receiv.	20	20	29	0.3	0.3
	LCO_2 prod.	300	240	48	4.4	3.6
	LCO_2 regas.	18	14	29	0.5	0.3
	Total	738	674		15.0	14.0
Chain C Pipeline	NG compr.	51	51	29	1.3	1.3
	CO_2 compr.	295	236	48	4.4	3.5
	Total	346	287		5.7	4.8

As can be seen from Table 2, the LEC requires roughly the same power as for pipeline transport (Chain C) of NG and CO_2. For long distances, however, ship transport including the LEC will be more effective than pipeline transport, where recompression is needed due to the frictional drop in the pipeline. When comparing transport chains for stranded NG, it is seen that the LEC requires about half of the total energy needed for ship transport of NG and CO_2 (Chain B). The LEC energy requirement is about the same as ship transport of NG without CO_2 capture (Chain A).

The required power is usually generated by conversion of NG to electricity or shaft work. There are several possible power processes with different efficiencies and applications e.g. open cycle gas turbines are often used offshore, whereas more efficient combined cycles can be used onshore. This will result in a larger NG fuel requirements for processes that demands power offshore. The exergy efficiency for power generation in an open cycle offshore is typically 29%, whereas an onshore power plant with CO_2 recovery has an exergy efficiency of 48% and a combined cycle power plant without CO_2 capture has an exergy efficiency of 55%. The chemical exergy for the NG is 14 057 kWh/tonne. Assuming the efficiencies given above, the loss of NG is 5.4% for the LEC, roughly the same as for pipeline transport, 4.8%. The LNG chains without (Chain A) and with (Chain B) CO_2 capture have about two and three times as large NG losses with 10.1% and 14.0%, respectively. Moreover, all the power required in the LEC is taken from a power plant with CO_2 capture, which means that most of the CO_2 will be captured. All other concepts will emit considerable amounts of CO_2 to the atmosphere.

Influence of Ambient Temperature and Feed Gas Pressure

Two parameters are especially important for the energy requirements; the feed gas inlet pressure and the ambient temperature. Therefore, a sensitivity analysis of these parameters is conducted, both for the LEC and for the conventional chains used in the benchmarking. It should also be noted that efficiency data for LNG transport are scarce and difficult to evaluate on an equal basis. However, a sensitivity analysis of the effect of temperature and inlet pressure on efficiency on the MFC process was conducted by Bauer et al.[10]. Data with references to inlet pressures and ambient temperatures are also found for the C3-PMR [9] and Niche-LNG [11]. Fig. 13 shows the energy requirements for the transport chains as a function of feed gas pressure and ambient temperature. A CO_2 capture and return rate of 85% is used in the calculations.

As can be seen from Fig. 13, the LEC has about the same efficiency as the novel offshore LNG process (Niche-LNG) from ABB without CO_2 sequestration. On the other hand, the energy requirements for the full scale LNG plants are lower than the LEC for pure LNG transport, and higher if the CO_2 transport is taken into account. As mentioned earlier, the efficiency of the offshore process is only 29% compared to 48% for a power plant onshore. Also, the CO_2 emitted from the open cycle power plant offshore will not be captured, in contrast to the onshore power plant, where CO_2 can be captured.

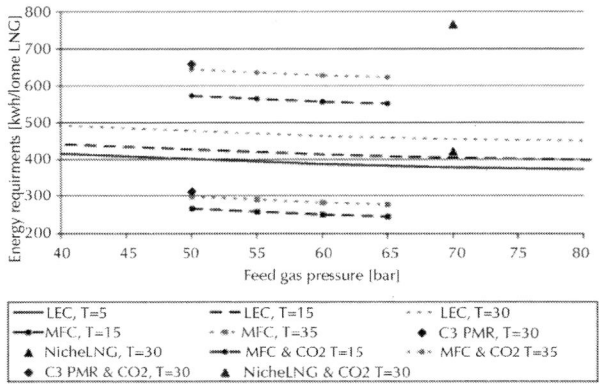

Figure 13: Energy requirements for the transport chains as a function of feed gas pressure and ambient temperature.

Fig. 14 shows the expected NG requirements for the LEC and benchmarking processes again as a function of feed gas pressure and ambient temperature. As can be seen, the LEC chain has the same NG requirements as the conventional LNG transport without CO_2 transport. If the CO_2 transport is included, the NG costs are doubled. Furthermore, the NG consumption for the Niche-LNG process including CO_2 transport is three times larger than for the LEC.

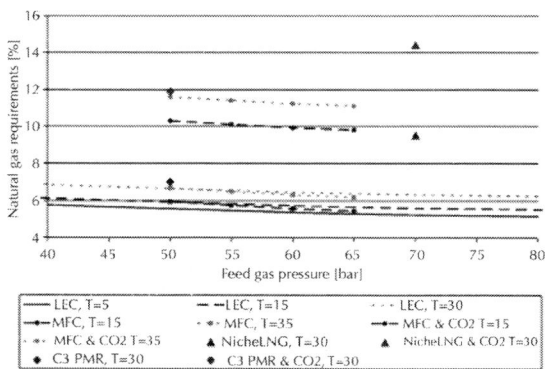

Figure 14: NG requirements for the transport chains as a function of feed gas pressure and ambient temperature.

The Complete Liquefied Energy Chain

The complete LEC includes a power process with CO_2 capture. Oxyfuel concepts are especially suited as they produce nitrogen as a by-product. Hence, the requirements for separate production of nitrogen (45 kWh/tonne LNG) is avoided, as the required amount of nitrogen is far less than what corresponds to the amount of oxygen needed in an oxyfuel power plant. In the LEC, NG at 70 bar is processed and transported from the field site to the market site where it is used for electricity production in an oxyfuel power plant. An efficiency of 47% for an oxyfuel process (with CO_2 capture) is reported in a study by Kvamsdal et al. [8]. However, the reported penalty of 3% points, based on LHV efficiency, for compression of CO_2 to 200 bar can be avoided. Therefore an LHV efficiency of 50% (which corresponds to an exergy efficiency of 48%) is used in the calculations. The CO_2 is conditioned and transported to the field site where it is unloaded at 150 bar. The exergy efficiency for the complete LEC is 46.4% as can be seen in Fig. 15.

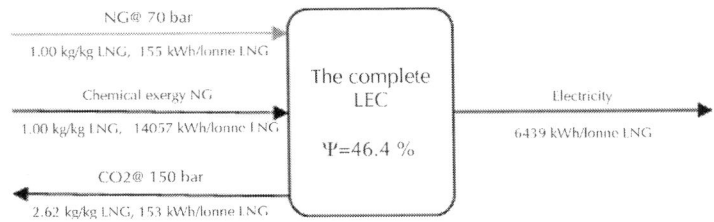

Figure 15: Battery limits and exergy efficiency for the full liquid energy chain.

A Comparison of the Complete LEC and Pipeline and Ship Transport Energy Chains for LNG and Co₂

In the calculations, a capture rate of 100% is used, although only 85% of the CO_2 is needed. This has minor implications on the process efficiencies, however, the ship utilization is slightly reduced. The complete LEC chain is compared with a conventional chain with and without CO_2 capture, as well as with pipeline transport. The results are shown in Table 3.

The complete LEC efficiency is 46.4%, whereas a similar conventional chain including production and transport of LNG and CO_2 and a power plant with CO_2 capture will have an efficiency of 42.0%. The conventional utilization of NG with LNG production and transport and a CC power plant without a CO_2 capture process will have an efficiency of 49.5%.

DISCUSSION

The exergy efficiency is highest when LNG is transported at a low pressure. However, the ship utilization is better for higher transport pressures, especially when all the produced CO_2 is returned to the field. The reason is that pressurized LNG requires less cryogenic cooling, hence less LIN is needed. In addition, the ship costs will increase for pressures higher than 6 bar. Hence, the optimum transport pressures should be determined taking into account both the onshore and offshore process as well as the ship to give the most energy and cost effective transport chain. An economic optimization is required to determine the optimal transport pressures. However, since the process efficiency also changes with transport pressure, the optimization must include the process models. Thus, the optimization of transport pressures should be done in an early design phase.

In conventional production of LNG, normally 6% of the NG is used in the gas turbines to produce power for the compressors and utilities. An offshore process needs to be compact and a loss of roughly 10% is expected. The recovered cryogenic exergy in LNG is 1.5% of the total chemical exergy giving an exergy efficiency for the offshore LNG process of 15%. Due to the cold water temperatures in the North Sea it may also be necessary to heat the CO_2 with hot water produced in a fuel gas heating system. As the exergy in pressurized CO_2 is only marginally larger than in liquid CO_2, the exergy efficiency for the heating and pumping of CO_2 is low, only 1.5%, meaning that almost all the provided exergy is lost, even when seawater is used for heating. The exergy efficiency in the offshore LEC process, which consists of an integrated process for liquefaction of LNG and heating and pumping of CO_2, is 87%. It is the extremely large improvement in this part of the chain that enables the LEC to be competitive even with conventional chainswithout CO_2 capture.

A conventional LNG chain without CCS will have an efficiency of 49.5%. The complete LEC efficiency is 46.4%, which is about the same as for a CCS chain with pipeline transport. A ship based CCS transport chain will have an efficiency of 42.0%. Stranded gas is referred to as gas that cannot be exploited economically by using pipelines. Therefore, for stranded NG, the total efficiency loss with CCS is 7.5%. By using the LEC these losses are reduced to 3.1%, which gives an extra power production of 6%. Although transport is generally regarded as the least technologically challenging part of the CCS chain, the large efficiency improvements shown here may by itself be larger than most proposals for savings in carbon capture power processes. Therefore the LEC is an important contribution to the technology development required to make a CCS chain economically attractive.

CONCLUSIONS

The liquid energy chain is an integrated transport chain for utilization of stranded NG for power production with CO_2 capture and use of CO_2 for EOR. By using LIN and LCO_2 as cold carriers, LNG can be produced offshore without extra power requirements and therefore no CO_2 emissions. It is found that the transport pressures will influence the efficiencies of the onshore and offshore process as well as the simple transport chain. The transport pressure will also influence the ship utilization. The thermodynamically optimal transport pressures are 5.5 bar for LCO_2, 6 bar for LIN and 1 bar for LNG. At these pressures, 1 kg of LNG requires 0.95 kg LIN and

2.21 kg LCO_2. Hence a 12 250 m³ ship will be able to transport 8330 m³ LNG to shore and requires 7180 m³ LCO_2 and 5070 m³ LIN from shore to the field, giving a ship utilization of 63.4% and a CO_2 sequestration rate of 85%. With these assumptions, the chain requires 319 kWh/tonne NG and gives a total efficiency of 46.4%; which is 4.4% points higher than a conventional transport chain for utilization of stranded NG with CCS. Compared to conventional ship based concepts the LEC will increase the power production with 6%, which is important both from a cost and resource perspective.

ACKNOWLEDGMENTS

The sponsors of the project are the Research Council of Norway, StatoilHydro and SINTEF Energy Research. Thanks to Jostein Pettersen and Ivar S. Ertesvåg for valuable discussions.

REFERENCES

1. Aspelund A, Gundersen T. A liquefied energy chain for transport and utilization of natural gas for power production with CO2 capture and storage – Part 1. Appl Energy 2009; 86(6):781–92.

2. Aspelund A, Gundersen T. A liquefied energy chain for transport and utilization of natural gas for power production with CO2 capture and storage – Part 2: the offshore and the onshore processes. Appl Energy 2009; 86(6):793–804.

3. Aspelund A, Tveit SP, Gundersen T. A liquefied energy chain for transport and utilization of natural gas for power production with CO2 capture and storage – Part 3: the combined carrier and onshore storage. Appl Energy 2009; 86(6):805–14.

4. NIST, National Institute of Standards and Technology, 2008. .

5. HYSYS. Aspen Tech, 2008. .

6. Kotas TJ. The exergy method of thermal plant analysis. Krieger; 1995.

7. Ertesvåg IS. Sensitivity of chemical exergy for atmospheric gases and gaseous fuels to variations in ambient conditions. J Energy Convers Manage 2007:48.

8. Kvamsdal HM, Jordal K, Bolland O. A quantitative comparison of gas turbine cycles with CO2 capture. J Energy 2007; 32:10–24.

9. Barclay M, Denton F. Selecting offshore LNG processes. J LNG 2005.

10. Bauer HC, Franke HE, Schier MT, Owren GA. MFCs3 the formula for high capacity LNG base load plants. In: AIChE spring national meeting, 4th topical conference on natural gas utilization, 2004, New Orleans, LA.

11. Festen L. Offshore LNG production. In: Lummus GPA conference, 2006, London.

12. Aspelund A, Sandvik TE, Wongraven LR, Krogstad H. Offshore unloading of CO2 to an oilfield. In: Seventh international conference on greenhouse gas control technologies, (GHGT-7), 2004, Vancouver.

13. Aspelund A, Mølnvik MJ, De Koeijer G. Ship transport of CO2 – technical solutions and analysis of costs energy utilization exergy efficiency and CO2 emissions. J Chem Eng Res Design 2006; 84:A9.

14. Aspelund A, Jordal K. Gas conditioning – the interface between CO2 capture and transport. J Greenhouse Gas Control 2007; 1:3.

15. Li H, Yan J, Yan J, Anheden M. Impurity impacts on purification process in oxyfuel combustion based CO2 capture and storage system. J Appl Energy 2009; 86(2):202–13.

16. Li H, Ji X, Yan J. A new modification on RK EOS for gaseous CO2 and gaseous mixtures of CO2 and H2O. Int J Energy Res 2006; 30:135–48.

17. Li H, Yan J. Evaluating cubic equations of state for calculation of vapor–liquid equilibrium of CO2 and CO2 mixtures for CO2 capture and storage processes. J Appl Energy 2009; 86(6):826–36.

Citations

CHAPTER 1

Sang Hyun Ahn, Insoo Choi, Oh Joong Kwon, Jae Jeong Kim, Hydrogen production through the fuel processing of liquefied natural gas with silicon-based micro-reactors, Chemical Engineering Journal, Volume 247, 1 July 2014, Pages 9-15, ISSN 1385-8947, http://dx.doi.org/10.1016/j.cej.2014.02.108.

CHAPTER 2

Zhixin Sun, Jiangfeng Wang, Yiping Dai, Jihong Wang, Exergy analysis and optimization of a hydrogen production process by a solar-liquefied natural gas hybrid driven transcritical CO2 power cycle, International Journal of Hydrogen Energy, Volume 37, Issue 24, December 2012, Pages 18731-18739, ISSN 0360-3199, http://dx.doi.org/10.1016/j.ijhydene.2012.08.028.

CHAPTER 3

Meysam Kamalinejad, Majid Amidpour, S.M. Mousavi Naeynian, Thermodynamic design of a cascade refrigeration system of liquefied natural gas by applying mixed integer non-linear programming, Chinese Journal of Chemical Engineering, Available online 3 February 2015, ISSN 1004-9541, http://dx.doi.org/10.1016/j.cjche.2014.05.023.

CHAPTER 4

Guanghui Xia, Qingxuan Sun, Xu Cao, Jiangfeng Wang, Yizhao Yu, Laisheng Wang, Thermodynamic analysis and optimization of a solar-powered transcritical CO2 (carbon dioxide) power cycle for reverse osmosis desalination based on the recovery of cryogenic energy of LNG (liquefied natural gas), Energy, Volume 66, 1 March 2014, Pages 643-653, ISSN 0360-5442, doi.org/10.1016/j.energy.2013.12.029.

CHAPTER 5

Yongju Bang, Jeong Gil Seo, Min Hye Youn, In Kyu Song, Hydrogen production by steam reforming of liquefied natural gas (LNG) over mesoporous Ni-Al2O3 aerogel catalyst prepared by a single-step epoxide-driven sol-gel method, International Journal of Hydrogen Energy, Volume 37, Issue 2, January 2012, Pages 1436-1443, ISSN 0360-3199, http://dx.doi.org/10.1016/j.ijhydene.2011.10.008.

CHAPTER 6

Jeong Gil Seo, Min Hye Youn, Ho-In Lee, Jae Jeong Kim, Eunsun Yang, Jin Suk Chung, Pil Kim, In Kyu Song, Hydrogen production by steam reforming of liquefied natural gas (LNG) over mesoporous nickel–alumina xerogel catalysts: Effect of nickel content, Chemical Engineering Journal, Volume 141, Issues 1–3, 15 July 2008, Pages 298-304, ISSN 1385-8947, http://dx.doi.org/10.1016/j.cej.2008.01.001.

CHAPTER 7

Alexandros Arsalis, Andreas Alexandrou, Thermoeconomic modeling and exergy analysis of a decentralized liquefied natural gas-fueled combined-cooling–heating-and-power plant, Journal of Natural Gas Science and Engineering, Volume 21, November 2014, Pages 209-220, ISSN 1875-5100, http://dx.doi.org/10.1016/j.jngse.2014.08.009.

CHAPTER 8

Audun Aspelund, Truls Gundersen, A liquefied energy chain for transport and utilization of natural gas for power production with CO2 capture and storage – Part 4: Sensitivity analysis of transport pressures and benchmarking with conventional technology for gas transport, Applied Energy, Volume 86, Issue 6, June 2009, Pages 815-825, ISSN 0306-2619, http://dx.doi.org/10.1016/j.apenergy.2008.10.02.

Index